BIG DATA BOOTCAMP

WHAT MANAGERS NEED TO KNOW TO PROFIT FROM THE BIG DATA REVOLUTION

David Feinleib

President and Publisher: Paul Manning
Acquisitions Editor: Jeff Olson
Editorial Board: Steve Anglin, Mark Beckner, Ewan Buckingham, Gary Cornell, Louise Corrigan, James DeWolf, Jonathan Gennick, Robert Hutchinson, Michelle Lowman, James Markham, Matthew Moodie, Jeff Olson, Jeffrey Pepper, Douglas Pundick, Ben Renow-Clarke, Dominic Shakeshaft, Gwenan Spearing, Matt Wade, Steve Weiss, Tom Welsh
Coordinating Editor: Rita Fernando
Copy Editor: Kezia Endsley
Compositor: SPi Global
Indexer: SPi Global
Cover Designer: Anna Ishchenko

Distributed to the book trade worldwide by Springer Science+Business Media New York, 233 Spring Street, 6th Floor, New York, NY 10013. Phone 1-800-SPRINGER, fax (201) 348-4505, e-mail orders-ny@springer-sbm.com, or visit www.springeronline.com. Apress Media, LLC is a California LLC and the sole member (owner) is Springer Science + Business Media Finance Inc (SSBM Finance Inc). SSBM Finance Inc is a Delaware corporation.

For information on translations, please e-mail rights@apress.com, or visit www.apress.com.

Apress and friends of ED books may be purchased in bulk for academic, corporate, or promotional use. eBook versions and licenses are also available for most titles. For more information, reference our Special Bulk Sales–eBook Licensing web page at www.apress.com/bulk-sales.

Any source code or other supplementary materials referenced by the author in this text is available to readers at www.apress.com. For detailed information about how to locate your book's source code, go to www.apress.com/source-code/.

Apress Business: The Unbiased Source of Business Information

Apress business books provide essential information and practical advice, each written for practitioners by recognized experts. Busy managers and professionals in all areas of the business world—and at all levels of technical sophistication—look to our books for the actionable ideas and tools they need to solve problems, update and enhance their professional skills, make their work lives easier, and capitalize on opportunity.

Whatever the topic on the business spectrum—entrepreneurship, finance, sales, marketing, management, regulation, information technology, among others—Apress has been praised for providing the objective information and unbiased advice you need to excel in your daily work life. Our authors have no axes to grind; they understand they have one job only—to deliver up-to-date, accurate information simply, concisely, and with deep insight that addresses the real needs of our readers.

It is increasingly hard to find information—whether in the news media, on the Internet, and now all too often in books—that is even-handed and has your best interests at heart. We therefore hope that you enjoy this book, which has been carefully crafted to meet our standards of quality and unbiased coverage.

We are always interested in your feedback or ideas for new titles. Perhaps you'd even like to write a book yourself. Whatever the case, reach out to us at editorial@apress.com and an editor will respond swiftly. Incidentally, at the back of this book, you will find a list of useful related titles. Please visit us at www.apress.com to sign up for newsletters and discounts on future purchases.

The Apress Business Team

Contents

About the Author. vii

Preface. ix

Introduction xi

Chapter 1:   Big Data. 1

Chapter 2:   The Big Data Landscape 15

Chapter 3:   Your Big Data Roadmap. 35

Chapter 4:   Big Data at Work 49

Chapter 5:   Why a Picture is Worth a Thousand Words. 63

Chapter 6:   The Intersection of Big Data, Mobile,
              and Cloud Computing 85

Chapter 7:   Doing a Big Data Project. 103

Chapter 8:   The Next Billion-Dollar IPO: Big
              Data Entrepreneurship. 125

Chapter 9:   Reach More Customers with Better
              Data—and Products 141

Chapter 10:   How Big Data Is Changing the Way We Live. 157

Chapter 11:   Big Data Opportunities in Education 173

Chapter 12:   Capstone Case Study: Big Data Meets Romance 189

Appendix A: Big Data Resources. 205

Index 209

About the Author

David Feinleib, is the producer of *The Big Data Landscape, Big Data Trends*, and *Big Data TV*, all of which may be found on the web at www.BigDataLandscape.com. Mr. Feinleib's Big Data Trends presentation was featured as "Hot On Twitter" and has been viewed more than 50,000 times on SlideShare. Mr. Feinleib has been quoted by *Business Insider* and *CNET*, and his writing has appeared on Forbes.com and in *Harvard Business Review China.* He is the Managing Director of The Big Data Group.

Prior to working at The Big Data Group, Mr. Feinleib was a general partner at Mohr Davidow Ventures. Mr. Feinleib co-founded Consera Software, which was acquired by HP; Likewise Software, which was acquired by EMC Isilon; and Speechpad, a leader in web-based audio-video transcription. He began his career at Microsoft. Mr. Feinleib holds a BA from Cornell University, graduating *summa cum laude*, and an MBA from the Graduate School of Business at Stanford University. The author of *Why Startups Fail* (Apress, 2011), he is an avid violinist and two-time Ironman finisher.

Preface

If it's March or December, watch out. You may be headed for a break up. Authors David McCandless and Lee Byron, two experts on data visualization, analyzed 10,000 Facebook status updates and plotted them on a graph. They figured out some amazing insights. Breakups spike around spring break and then again two weeks before the winter holidays.

If it's Christmas Day, on the other hand, you're in good shape. Fewer breakups happen on Christmas than on any other day of the year. If you're thinking that Big Data is a far off topic with little relevance to your daily life, think again. Data is changing how dating sites organize user profiles, how marketers target you to get you to buy products, and even how we track our fitness goals so we can lose weight.

My own obsession with Big Data began while I was training for Ironman France. I started tracking every hill I climbed, every mile I ran, and every swim I completed in the icy cold waters of San Francisco's Aquatic Park. Then I uploaded all that information to the web so that I could review it, visualize it, and analyze it. I didn't know it at the time, but that was the start of a fascinating exploration into what is now known as Big Data.

Airlines and banks have used data for years to figure out what price to charge and who to give loans to. Credit card companies use data to detect fraud. But it wasn't until relatively recently that data—Big Data as it is talked about today—really became a part of our daily lives. That's because even though these companies worked with lots of data, that data was more or less invisible to us.

Then came Facebook and Google and the data game changed forever. You and I and every other user of those services generate a data trail that reflects our behavior. Every time we search for something, "Like" someone, or even just visit a web page, we add to that trail. When Facebook had just a few users, storing all that data about what we were doing was no big deal. But existing technologies soon became unable to meet the needs of a trillion web searches and more than a billion friends.

These companies had to build new technologies for them to store and analyze data. The result was an explosion of innovation called Big Data. Other companies saw what Facebook and Google were doing and wanted to make use of data in the same way to figure out what we wanted to buy so they could sell us more of their products. Entrepreneurs wanted to use that data to offer better access to healthcare. Municipal governments wanted to use it to understand the residents of their cities better and determine what services to provide.

But a huge problem remained.

Most companies have lots of data. But most employees are not data scientists. As a result, the conversation around Big Data remained far too technical to be accessible to a broad audience.

There was an opportunity to take a heavily technical subject—one that had a relatively geeky bent to it—and open it up to everyone, to explain the impact of data on our daily lives. This book is the result. It is the story of how data is changing not only the way we work but also the way we live, love, and learn.

As with any major undertaking, many people provided assistance and support, for which I am deeply grateful. I would particularly like to thank Yuchi Chou, without whom this book would not exist; Jeff Olson at Apress for spearheading this project and for his and his team's incredible editing skills; Cameron Myhrvold for his enduring mentorship; Houston Jayne at Walmart.com and Scott Morell at AT&T, and their respective teams, for their insights and support; and Jon Feiber, Mark Gorenberg, and Joe and Nancy Schoendorf for their wisdom and advice.

—David Feinleib

San Francisco, California

September, 2014

Introduction

Although earthquakes have been happening for millions of years and we have lots of data about them, we still can't predict exactly when and where they'll happen. Thousands of people die every year as a result and the costs of material damage from a single earthquake can run into the hundreds of billions of dollars.

The problem is that based on the data we have, earthquakes and almost-earthquakes look roughly the same, right up until the moment when an almost-earthquake becomes the real thing. But by then, of course, it's too late.

And if scientists were to warn people every time they thought they recognized the data for what appeared to be an earthquake, there would be a lot of false-alarm evacuations. What's more, much like the boy who cried wolf, people would eventually tire of false alarms and decide not to evacuate, leaving them in danger when the real event happened.

When Good Predictions Aren't Good Enough

To make a *good* prediction, therefore, a few things need to be true. We must have enough data about the past to identify patterns. The events associated with those patterns have to happen consistently. And we have to be able to differentiate what looks like an event but isn't from an actual event. This is known as ruling out false positives.

But a good prediction alone isn't enough to be useful. For a prediction to be *useful*, we have to be able to act on a prediction early enough and fast enough for it to matter.

When a real earthquake is happening, the data very clearly indicates as much. The ground shakes, the earth moves, and, once the event is far enough along, the power goes out, explosions occur, poisonous gas escapes, and fires erupt. By that time, of course, it doesn't take a lot of computers or talented scientists to figure out that something bad is happening.

So to be useful, the data that represents the present needs to look like that of the past far enough in advance for us to act on it. If we can only make the match a few seconds before the actual earthquake, it doesn't matter. We need sufficient time to get the word out, mobilize help, and evacuate people.

What's more, we need to be able to perform the analysis of the data itself fast enough to matter. Suppose we had data that could tell us a day in advance that an earthquake was going to happen. If it takes us two days to analyze that data, the data and our resulting prediction wouldn't matter.

This at its core is both the challenge and the opportunity of Big Data. Just having data isn't enough. We need relevant data early enough and we have to be able to analyze it fast enough that we have sufficient time to act on it. The sooner an event is going to happen, the faster we need to be able to make an accurate prediction. But at some point we hit the law of diminishing returns. Even if we can analyze immense amounts of data in seconds to predict an earthquake, such analysis doesn't matter if there's not enough time left to get people out of harm's way.

Enter Big Data: Speedier Warnings and Lives Saved

On October 22, 2012, six engineers were sentenced to six-year jail sentences after being accused of inappropriately reassuring villagers about a possible upcoming earthquake. The earthquake occurred in 2009 in the town of L'Aquila, Italy; 300 villagers died.

Could Big Data have helped the geologists make better predictions?

Every year, some 7,000 earthquakes occur around the world of magnitude 4.0 or greater. Earthquakes are measured either on the well-known Richter scale, which assigns a number to the energy contained in an earthquake, or the more recent moment magnitude scale (MMS), which measures an earthquake in terms of the amount of energy released.[1]

When it comes to predicting earthquakes, there are three key questions that must be answered: when, where, and how big? In[2] *The Charlatan Game*, Matthew A. Mabey of Brigham Young University argues that while there are precursors to earthquakes, "we can't yet use them to reliably or usefully predict earthquakes."

---

[1] http://www.gps.caltech.edu/uploads/File/People/kanamori/HKjgr79d.pdf
[2] http://www.dnr.wa.gov/Publications/ger_washington_geology_2001_v28_no3.pdf

Instead, the best we can do is prepare for earthquakes, which happen a lot more often than people realize. Preparation means building bridges and buildings that are designed with earthquakes in mind and getting emergency kits together so that infrastructure and people are better prepared when a large earthquake strikes.

Earthquakes, as we all learned back in our grade school days, are caused by the rubbing together of tectonic plates—those pieces of the Earth that shift around from time to time.

Not only does such rubbing happen far below the Earth's surface, but the interactions of the plates are complex. As a result, good earthquake data is hard to come by, and understanding what activity causes what earthquake results is virtually impossible.[3]

Ultimately, accurately predicting earthquakes—answering the questions of when, where, and how big—will require much better data about the natural elements that cause earthquakes to occur and their complex interactions.

Therein lies a critical lesson about Big Data: predictions are different than forecasts. Scientists can *forecast* earthquakes but they cannot predict them. When will San Francisco experience another quake like that of 1906, which resulted in more than 3,000 casualties? Scientists can't say for sure.

They can forecast the probability that a quake of a certain magnitude will happen in a certain region in a certain time period. They can say, for example, that there is an 80% likelihood that a magnitude 8.4 earthquake will happen in the San Francisco Bay Area in the next 30 years. But they cannot say when, where, and how big that earthquake will happen with complete certainty. Thus the difference between a forecast and a prediction.[4]

But if there is a silver lining in the ugly cloud that is earthquake forecasting, it is that while earthquake prediction is still a long way off, scientists are getting smarter about buying potential earthquake victims a few more seconds. For that we have Big Data methods to thank.

Unlike traditional earthquake sensors, which can cost $3,000 or more, basic earthquake detection can now be done using low-cost sensors that attach to standard computers or even using the motion sensing capabilities built into many of today's mobile devices for navigation and game-playing.[5]

---

[3]http://www.planet-science.com/categories/over-11s/natural-world/2011/03/can-we-predict-earthquakes.aspx

[4]http://ajw.asahi.com/article/globe/feature/earthquake/AJ201207220049

[5]http://news.stanford.edu/news/2012/march/quake-catcher-warning-030612.html

The Stanford University Quake-Catcher Network (QCN) comprises the computers of some 2,000 volunteers who participate in the program's distributed earthquake detection network. In some cases, the network can provide up to 10 seconds of early notification to those about to be impacted by an earthquake. While that may not seem like a lot, it can mean the difference between being in a moving elevator or a stationary one or being out in the open versus under a desk.

The QCN is a great example of the kinds of low-cost sensor networks that are generating vast quantities of data. In the past, capturing and storing such data would have been prohibitively expensive. But, as we will talk about in future chapters, recent technology advances have made the capture and storage of such data significantly cheaper—in some cases more than a hundred times cheaper than in the past.

Having access to both more and better data doesn't just present the possibility for computers to make smarter decisions. It lets humans become smarter too. We'll find out how in just a moment—but first let's take a look at how we got here.

Big Data Overview

When it comes to Big Data, it's not how much data we have that really matters, but what we do with that data.

Historically, much of the talk about Big Data has centered around the three Vs—volume, velocity and variety. Volume refers to the quantity of data you're[6] working with. Velocity means how quickly that data is flowing. Variety refers to the diversity of data that you're working with, such as marketing data combined with financial data, or patient data combined with medical research and environmental data.

But the most important "V" of all is value. The real measure of Big Data is not its size but rather the scale of its impact—the value Big Data that delivers to your business or personal life. Data for data's sake serves very little purpose. But data that has a positive and outsized impact on our business or personal lives truly is Big Data.

When it comes to Big Data, we're generating more and more data every day. From the mobile phones we carry with us to the airplanes we fly in, today's systems are creating more data than ever before. The software that operates these systems gathers immense amounts of data about what these systems are doing and how they are performing in the process. We refer to these measurements as event data and the software approach for gathering that data as instrumentation.

---

[6]This definition was first proposed by industry analyst Doug Laney in 2001.

For example, in the case of a web site that processes financial transactions, instrumentation allows us to monitor not only how quickly users can access the web site, but also the speed at which the site can read information from a database, the amount of memory consumed at any given time by the servers the site is running on, and, of course, the kinds of transactions users are conducting on the site. By analyzing this stream of event data, software developers can dramatically improve response time, which has a significant impact on whether users and customers remain on a web site or abandon it.

In the case of web sites that handle financial or commerce transactions, developers can also use this kind of event stream data to reduce fraud by looking for patterns in how clients use the web site and detecting unusual behavior. Big Data-driven insights like these lead to more transactions processed and higher customer satisfaction.

Big Data provides insights into the behavior of complex systems in the real world as well. For example, an airplane manufacturer like Boeing can measure not only internal metrics such as engine fuel consumption and wing performance but also external metrics like air temperature and wind speed.

This is an example of how quite often the value in Big Data comes not from one data source by itself, but from bringing multiple data sources together. Data about wind speed alone might not be all that useful. But bringing data about wind speed, fuel consumption, and wing performance together can lead to new insights, resulting in better plane designs. These in turn provide greater comfort for passengers and improved fuel efficiency, resulting in lower operating costs for airlines.

When it comes to our personal lives, instrumentation can lead to greater insights about an altogether different complex system—the human body. Historically, it has often been expensive and cumbersome for doctors to monitor patient health and for us as individuals to monitor our own health. But now, three trends have come together to reduce the cost of gathering and analyzing health data.

These key trends are the widespread adoption of low-cost mobile devices that can be used for measurement and monitoring, the emergence of cloud-based applications to analyze the data these devices generate, and of course the Big Data itself, which in combination with the right analytics software and services can provide us with tremendous insights. As a result, Big Data is transforming personal health and medicine.

Big Data has the potential to have a positive impact on many other areas of our lives as well, from enabling us to learn faster to helping us stay in the relationships we care about longer. And as we'll learn, Big Data doesn't just make computers smarter—it makes human beings smarter too.

How Data Makes Us Smarter

If you've ever wished you were smarter, you're not alone. The good news, according to recent studies, is that you can actually increase the size of your brain by adding more data.

To become licensed to drive, London cab drivers have to pass a test known somewhat ominously as "the Knowledge," demonstrating that they know the layout of downtown London's 25,000 streets as well as the location of some 20,000 landmarks. This task frequently takes three to four years to complete, if applicants are able to complete it at all. So do these cab drivers actually get smarter over the course of learning the data that comprises the Knowledge?[7]

It turns out that they do.

Data and the Brain

Scientists once thought that the human brain was a fixed size. But brains are "plastic" in nature and can change over time, according to a study by Professor Eleanor Maguire of the Wellcome Trust Centre for Neuroimaging at University College London.[8]

The study tracked the progress of 79 cab drivers, only 39 of whom ultimately passed the test. While drivers cited many reasons for not passing, such as a lack of time and money, certainly the difficulty of learning such an enormous body of information was one key factor. According to the City of London web site, there are just 25,000 licensed cab drivers in total, or about one cab driver for every street.[9]

After learning the city's streets for years, drivers evaluated in the study showed "increased gray matter" in an area of the brain called the posterior hippocampus. In other words, the drivers actually grew more cells in order to store the necessary data, making them smarter as a result.

Now, these improvements in memory did not come without a cost. It was harder for drivers with expanded hippocampi to absorb new routes and to form new associations for retaining visual information, according to another study by Maguire.[10]

---

[7]http://www.tfl.gov.uk/businessandpartners/taxisandprivatehire/1412.aspx
[8]http://www.scientificamerican.com/article.cfm?id=london-taxi-memory
[9]http://www.tfl.gov.uk/corporate/modesoftransport/7311.aspx
[10]http://www.ncbi.nlm.nih.gov/pubmed/19171158

Similarly, in computers, advantages in one area also come at a cost to other areas. Storing a lot of data can mean that it takes longer to process that data. Storing less data may produce faster results, but those results may be less informed.

Take for example the case of a computer program trying to analyze historical sales data about merchandise sold at a store so it can make predictions about sales that may happen in the future.

If the program only had access to quarterly sales data, it would likely be able to process that data quickly, but the data might not be detailed enough to offer any real insights. Store managers might know that certain products are in higher demand during certain times of the year, but they wouldn't be able to make pricing or layout decisions that would impact hourly or daily sales.

Conversely, if the program tried to analyze historical sales data tracked on a minute-by-minute basis, it would have much more granular data that could generate better insights, but such insights might take more time to produce. For example, due to the volume of data, the program might not be able to process all the data at once. Instead, it might have to analyze one chunk of it at a time.

Big Data Makes Computers Smarter and More Efficient

One of the amazing things about licensed London cab drivers is that they're able to store the entire map of London, within six miles of Charing Cross, in memory, instead of having to refer to a physical map or use a GPS.

Looking at a map wouldn't be a problem for a London cab driver if the driver didn't have to keep his eye on the road and hands on the steering wheel, and if he didn't also have to make navigation decisions quickly. In a slower world, a driver could perhaps plot out a route at the start of a journey, then stop and make adjustments along the way as necessary.

The problem is that in London's crowded streets no driver has the luxury to perform such slow calculations and recalculations. As a result, the driver has to store the whole map in memory. Computer systems that must deliver results based on processing large amounts of data do much the same thing: they store all the data in one storage system, sometimes all in memory, sometimes distributed across many different physical systems. We'll talk more about that and other approaches to analyzing data quickly in the chapters ahead.

Fortunately if you want a bigger brain, memorizing the London city map isn't the only way to increase the size of your hippocampus. The good news, according to another study, is that exercise can also make your brain bigger.[11]

As we age, our brains shrink, leading to memory impairment. According to the authors of the study, who did a trial with 120 older adults, exercise training increased the size of the hippocampal volume of these adults by 2%, which was associated with improved memory function. In other words, keeping sufficient blood flowing through our brains can help prevent us from getting dumber. So if you want to stay smart, work out.

Unlike humans, however, computers can't just go to the gym to increase the size of their memory. When it comes to computers and memory, there are three options: add more memory, swap data in and out of memory, or compress the data.

A lot of data is redundant. Just think of the last time you wrote a sentence or multiplied some large numbers together. Computers can save a lot of space by compressing repeated characters, words, or even entire phrases in much the same way that court reporters use shorthand so they don't have to type every word.

Adding more memory is expensive, and typically the faster the memory, the more expensive it is. According to one source, Random Access Memory or RAM is 100,000 times faster than disk memory. But it is also about 100 times more expensive.[12]

It's not just the memory itself that costs so much. More memory comes with other costs as well.

There are only so many memory chips that can fit in a typical computer, and each memory stick can hold a certain number of chips. Power and cooling are issues too. More electronics require more electricity and more electricity generates more heat. Heat needs to be dissipated or cooled, which in and of itself requires more electricity (and generates more heat). All of these factors together make the seemingly simple task of adding more memory a fairly complex one.

Alternatively, computers can just use the memory they have available and swap the needed information in and out. Instead of trying to look at all available data about car accidents or stock prices at once, for example, a computer can load yesterday's data, then replace that with data from the day before, and so on. The problem with such an approach is that if you're looking for patterns that span multiple days, weeks, or years, swapping all that data in and out takes a lot of time and makes those patterns hard to find.

---

[11]http://www.pnas.org/content/early/2011/01/25/1015950108.full.pdf
[12]http://research.microsoft.com/pubs/68636/ms_tr_99_100_rules_of_thumb_in_data_engineering.pdf

In contrast to machines, human beings don't require a lot more energy to use more brainpower. According to an article in Scientific American, the brain "continuously slurps up huge amounts of energy."[13]

But all that energy is remarkably small compared to that required by computers. According to the same article, "a typical adult human brain runs on around 12 watts—a fifth of the power required by a standard 60 watt light bulb." In contrast, "IBM's Watson, the supercomputer that defeated Jeopardy! champions, depends on ninety IBM Power 750 servers, each of which requires around one thousand watts." What's more, each server weighs about 120 pounds.

When it comes to Big Data, one challenge is to make computers smarter. But another challenge is to make them more efficient.

On February 16, 2011, a computer created by IBM known as Watson beat two *Jeopardy!* champions to win $77,147. Actually, Watson took home $1 million in prize money for winning the epic man versus machine battle. But was Watson really smart in the way that the other two contestants on the show were? Can Watson think for itself?

With an estimated $30 million in research and development investment, 200 million pages of stored content, and some 2,800 processor cores, there's no doubt that Watson is very good at answering *Jeopardy!* questions.

But it's difficult to argue that Watson is intelligent in the way that, say, HAL was in the movie *2001: A Space Odyssey*. And Watson isn't likely to express its dry humor like one of the show's other contestants, Ken Jennings, who wrote "I for one welcome our new computer overlords," alongside his final Jeopardy! answer. What's more, Watson can't understand human speech; rather, the computer is restricted to processing Jeopardy! answers in the form of written text.

Why can't Watson understand speech? Watson's designers felt that creating a computer system that could come up with correct Jeopardy! questions was hard enough. Introducing the problem of understanding human speech would have added an extra layer of complexity. And that layer is a very complex one indeed.

Although there have been significant advances in understanding human speech, the solution is nowhere near flawless. That's because, as Markus Forsberg at the Chalmers Institute of Technology highlights, understanding human speech is no simple matter.[14]

---

[13]http://www.scientificamerican.com/article.cfm?id=thinking-hard-calories
[14]http://www.speech.kth.se/~rolf/gslt_papers/MarkusForsberg.pdf

Speech would seem to fit at least some of the requirements for Big Data. There's a lot of it and by analyzing it, computers should be able to create patterns for recognizing it when they see it again. But computers face many challenges in trying to understand speech.

As Forsberg points out, we use not only the actual sound of speech to understand it but also an immense amount of contextual knowledge. Although the words "two" and "too" sound alike, they have very different meanings. This is just the start of the complexity of understanding speech. Other issues are the variable speeds at which we speak, accents, background noise, and the continuous nature of speech—we don't pause between each word, so trying to convert individual words into text is an insufficient approach to the speech recognition problem.

Even trying to group words together can be difficult. Consider the following examples cited by Forsberg:

- It's not easy to wreck a nice beach.

- It's not easy to recognize speech.

- It's not easy to wreck an ice beach.

Such sentences sound very similar yet at the same time very different.

But computers are making gains, due to a combination of the power and speed of modern computers, combined with advanced new pattern-recognition approaches. The head of Microsoft's[15] research and development organization stated that the company's most recent speech recognition technology is 30% more accurate than the previous version—meaning that instead of getting one out of every four or five words wrong, the software gets only one out of every seven or eight incorrect. Pattern recognition is also being used for tasks like machine-based translation—but as users of Google Translate will attest, these technologies still have a long way to go.

Likewise, computers are still far off from being able to create original works of content, although, somewhat amusingly, people have tried to get them to do so. In one recent experiment, a programmer created a series of virtual programs to simulate monkeys typing randomly on keyboards, with the goal of answering the classic question of whether monkeys could recreate the works of William Shakespeare.[16] The effort failed, of course.

But computers are getting smarter. So smart, in fact, that they can now drive themselves.

---

[15]http://www.nytimes.com/2012/11/24/science/scientists-see-advances-in-deep-learning-a-part-of-artificial-intelligence.html?pagewanted=2&_r=0
[16]http://www.bbc.co.uk/news/technology-15060310

How Big Data Helps Cars Drive Themselves

If you've used the Internet, you've probably used Google Maps. The company, well known for its market dominating search engine, has accumulated more than 20 petabytes of data for Google Maps. To put that in perspective, it would take more than 82,000 256 GB hard drives of a typical Apple MacBook Pro computer to store all that data.[17]

But does all that data really translate into cars that can drive themselves? In fact, it does. In an audacious project to build self-driving cars, Google combines a variety of mapping data with information from a real-time laser detection system, multiple radars, GPS, and other devices that allow the system to "see" traffic, traffic lights, and roads, according to Sebastian Thrun, a Stanford University professor who leads the project at Google.[18]

Self-driving cars not only hold the promise of making roads safer, but also of making them more efficient by better utilizing the vast amount of empty space between cars on the road. According to one source, some 43,000 people in the United States die each year from car accidents and there are some five and a quarter million accidents per year in total.[19]

Google Cars can't think for themselves, per se, but they can do a great job at pattern matching. By combining existing data from maps with real-time data from a car's sensors, the cars can make driving decisions. For example, by matching against a database of what different traffic lights look like, self-driving cars can determine when to start and stop.

All of this would not be possible, of course, without three key elements that are a common theme of Big Data. First, the computer systems in the cars have access to an enormous amount of data. Second, the cars make use of sensors that take in all kinds of real-time information about the position of other cars, obstacles, traffic lights, and terrain. While these sensors are expensive today— the total cost of equipment for a self-driving equipped car is approximately $150,000—the sensors are expected to decrease in cost rapidly.

Finally, the cars can process all that data at a very high speed and make corresponding real-time decisions about what to do next as a result—all with a little computer equipment and a lot of software in the back seat.

---

[17]http://mashable.com/2012/08/22/google-maps-facts/
[18]http://spectrum.ieee.org/automaton/robotics/artificial-intelligence/how-google-self-driving-car-works
[19]http://www.usacoverage.com/auto-insurance/how-many-driving-accidents-occur-each-year.html

To put that in perspective, consider that just a little over 60 years ago, the UNIVAC computer, known for successfully predicting the results of the Eisenhower presidential election, took up as much space as a single car garage.[20]

How Big Data Enables Computers to Detect Fraud

All of this goes to show that computers are very good at performing high-speed pattern matching. That's a very useful ability not just on the road but off the road as well. When it comes to detecting fraud, fast pattern matching is critical.

We've all gotten that dreaded call from the fraud-prevention department of our credit card company. The news is never good—the company believes our credit card information has been stolen and that someone else is buying things at the local hardware store in our name. The only problem is that the local hardware store in question is 5,000 miles away.

Computers that can process greater amounts of data at the same time can make better decisions, decisions that have an impact on our daily lives. Consider the last time you bought something with your credit card online, for example.

When you clicked that Submit button, the action of the web site charging your card triggered a series of events. The proposed transaction was sent to computers running a complex set of algorithms used to determine whether you were you or whether someone was trying to use your credit card fraudulently.

The trouble is that figuring out whether someone is a fraudster or who they really claim to be is a hard problem. With so many data breaches and so much personal information available online, it's often the case that fraudsters know almost as much about you as you do.

Computer systems detect whether you are who you say you are in a few basic ways. They verify information. When you call into your bank and they ask for your name, address, and mother's maiden name, they compare the information you give them with the information they have on file. They may also look at the number you're calling from and see if it matches the number they have for you on file. If those pieces of information match, it's likely that you are who you say you are.

Computer systems also evaluate a set of data points about you to see if those seem to verify you are who you say you are or reduce that likelihood. The systems produce a confidence score based on the data points.

For example, if you live in Los Angeles and you're calling in from Los Angeles, that might increase the confidence score. However, if you reside in Los Angeles and are calling from Toronto, that might reduce the score.

More advanced scoring mechanisms (called algorithms) compare data about you to data about fraudsters. If a caller has a lot of data points in common with fraudsters, that might indicate that someone is a fraudster.

If the user of a web site is connecting from a computer other than the one they've connected from in the past, they have an out-of-country location (say Russia when they typically log in from the United States), and they've attempted a few different passwords, that could be indicative of a fraudster. The computer system compares all of these identifiers to common patterns of behavior for fraudsters and common patterns of behavior for you, the user, to see whether the identity confidence score should go up or down.

Lots of matches with fraudster patterns or differences from your usual behavior and the score goes down. Lots of matches with your usual behavior and the score goes up.

The problem for computers, however, is two-fold. First, they need a lot of data to figure out what your usual behavior is and what the behavior of a fraudster is. Second, once the computer knows those things, it has to be able to compare your behavior to these patterns while also performing that task for millions of other customers at the same time.

So when it comes to data, computers can get smarter in two ways. Their algorithms for detecting normal and abnormal behavior can improve and the amount of data they can process at the same time can increase.

What really puts both computers and cab drivers to the test, therefore, is the need to make decisions quickly. The London cab driver, like the self-driving car, has to know which way to turn and make second-by-second decisions depending on traffic and other conditions. Similarly, the fraud-detection program has to decide whether to approve or deny your transaction in a matter of seconds.

As Robin Gilthorpe, former CEO of Terracotta, a technology company, put it, "no one wants to be the source of a 'no,' especially when it comes to e-commerce."[21] A denied transaction to a legitimate customer means not only a lost sale but an unhappy customer. And yet denying fraudulent transactions is the key to making non-fraudulent transactions work.

---

[21]Briefing with Robin Gilthorpe, October 30, 2012.

Peer-to-peer payments company PayPal found that out firsthand when the company had to build technology early on to combat fraudsters, as early PayPal analytics expert Mike Greenfield has pointed out. Without such technology, the company would not have survived and people wouldn't have been able to make purchases and send money to each other as easily as they were able to.[22]

Better Decisions Through Big Data

As with any new technology, Big Data is not without its risks. Data in the wrong hands can be used for malicious purposes, and bad data can lead to bad decisions. As we continue to generate more data and as the software we use to analyze that data becomes more sophisticated, we must also become more sophisticated in how we manage and use the data and the insights we generate. Big Data is no substitute for good judgment.

When it comes to Big Data, human beings can still make bad decisions—such as running a red light, taking a wrong turn, or drawing a bad conclusion. But as we've seen here, we have the potential, through behavioral changes, to make ourselves smarter. We've also seen that technology can help us be more efficient and make fewer mistakes—the self-driving car, for example, can help us avoid driving through that red light or taking a wrong turn. In fact, over the next few decades, such technology has the potential to transform the entire transportation industry.

When it comes to making computers smarter, that is, enabling computers to make better decisions and predictions, what we've seen is that there are three main factors that come into play: data, algorithms, and speed.

Without *enough* data, it's hard to recognize patterns. Enough data doesn't just mean having all the data. It means being able to run analysis on enough of that data at the same time to create algorithms that can detect patterns. It means being able to test the results of the analysis to see if our conclusions are correct. Sampling one day of data might be useless, but sampling 10 years of data might produce results.

At the same time, all the data in the world doesn't mean anything if we can't process it fast enough. If you have to wait 10 minutes while standing in the grocery line for a fraud-detection algorithm to determine whether you can use your credit card, you're not likely to use that credit card for much longer. Similarly, if self-driving cars can only go at a snail's pace because they need more time to figure out whether to stop or move forward, no one will adopt self-driving cars. So speed plays a critical role as well when it comes to Big Data.

---

[22]http://numeratechoir.com/2012/05/

We've also seen that computers are incredibly efficient at some tasks, such as detecting fraud by rapidly analyzing vast quantities of similar transactions. But they are still inefficient relative to human beings at other tasks, such as trying to convert the spoken word into text. That, as we'll explore in the chapters ahead, constitutes one of the biggest opportunities in Big Data, an area called unstructured data.

Roadmap of the Book

In *Big Data Bootcamp*, we'll explore a range of different topics related to Big Data. In Chapter 1, we'll look at what Big Data is and how big companies like Amazon, Facebook, and Google are putting Big Data to work. We'll explore the dramatic shift in information technology, in which competitive advantage is coming less and less from technology itself than from information that is enabled by technology. We'll also dive into Big Data Applications (BDAs) and see how companies no longer need to build as much themselves and can instead rely on off-the-shelf applications to meet their Big Data needs, while they focus on the business problems they want to solve.

In Chapter 2, we'll look at the Big Data Landscape in detail. Originally a way for me to map out the Big Data space, the Big Data Landscape has become an entity in its own right, now used as an industry and government reference. We'll look at where venture capital investments are going and where exciting new companies are emerging to make Big Data ever more accessible to a wider audience.

Chapters 3, 4, and 5 explore Big Data from a few different angles. First, we'll lay the groundwork in Chapter 3 as we cover how to create your own Big Data roadmap. We'll look at how to choose new technologies and how to work with the ones you've already got—as well as at the emerging role of the chief data officer.

In Chapter 4 we'll explore the intersection of Big Data and design and how leading companies like Apple and Facebook find the right balance between relying on data and intuition in designing new products. In Chapter 5, we'll cover data visualization and the powerful ways in which it can make complex data sets easy to understand. We'll also cover some popular tools, readily available public data sets, and how you can get started creating your own visualizations in the cloud or on your desktop.

Starting in Chapter 6, we look at the all-important intersection of Big Data, mobile, and cloud computing and how these technologies are coming together to disrupt multiple billion-dollar industries. You'll learn what you need to know to transform your own with cloud, mobile, and Big Data capabilities.

In Chapter 7, we'll go into detail about how to do your own Big Data project. We'll cover the resources you need, the cloud technologies available, and who you'll need on your team to accomplish your Big Data goals. We'll cover three real-world case studies: churn reduction, marketing analytics, and the connected car. These critical lessons can be applied to nearly any Big Data business problem.

Building on everything we've learned about Big Data, we'll jump back into the business of Big Data in Chapter 8, where we explore opportunities for new businesses that take advantage of the Big Data opportunity. We'll also look at the disruptive subscription and cloud-based delivery models of Software as a Service (SaaS) and how to apply it to your Big Data endeavors. In Chapter 9, we'll look at Big Data from the marketing perspective—how you can apply Big Data to reach and interact with customers more effectively.

Finally, in chapters 10, 11, and 12 we'll explore how Big Data touches not just our business lives but our personal lives as well, in the areas of health and well-being, education, and relationships. We'll cover not only some of the exciting new Big Data applications in these areas but also the many opportunities to create new businesses, applications, and products.

I look forward to joining you on the journey as we explore the fascinating topic of Big Data together. I hope you will enjoy reading about the tremendous Big Data opportunities available to you as much as I enjoy writing about them.

Big Data

What It Is, and Why You Should Care

Scour the Internet and you'll find dozens of definitions of Big Data. There are the three v's—volume, variety, and velocity. And there are the more technical definitions, like this one from Edd Dumbill, analyst at O'Reilly Media: "Big Data is data that exceeds the processing capacity of conventional database systems. The data is too big, moves too fast, or doesn't fit the strictures of your database architectures. To gain value from this data, you must choose an alternative way to process it."[1]

Such definitions, while accurate, miss the true value of Big Data. Big Data should be measured by the size of its impact, not by the amount of storage space or processing power that it consumes. All too often, the discussion around Big Data gets bogged down in terabytes and petabytes, and in how to store and process the data rather than in how to use it.

As consumers and business users, the size and scale of data isn't what we care about. Rather, we want to be able to ask and answer the questions that matter to us. What medicine should we take to address a serious health condition? What information, study tools, and exercises should we give students to help them learn more effectively? How much more should we spend on a marketing campaign? Which features of a new product are our customers using?

That is what Big Data is really all about. It is the ability to capture and analyze data and gain actionable insights from that data at a much lower cost than was historically possible.

---

[1] http://radar.oreilly.com/2012/01/what-is-big-data.html

What is truly transformative about Big Data is the ease with which we can now use data. No longer do we need complex software that takes months or years to set up and use. Nearly all the analytics power we need is available through simple software downloads or in the cloud.

No longer do we need expensive devices to collect data. Now we can collect performance and driving data from our cars, fitness and location data from GPS watches, and even personal health data from low-cost attachments to our mobile phones. It is the combination of these capabilities—Big Data meets the cloud meets mobile—that is truly changing the game when it comes to making it easy to use and apply data.

■ **Note** Big Data is transformative: You don't need complex software or expensive data-collection techniques to make use of it. Big Data meeting the cloud and mobile worlds is a game changer for businesses of all sizes.

Big Data Crosses Over Into the Mainstream

So why has Big Data become so hot all of a sudden? Big Data has broken into the mainstream due to three trends coming together.

First, multiple high-profile consumer companies have ramped up their use of Big Data. Social networking behemoth Facebook uses Big Data to track user behavior across its network. The company makes new friend recommendations by figuring out who else you know.

The more friends you have, the more likely you are to stay engaged on Facebook. More friends means you view more content, share more photos, and post more status updates.

Business networking site LinkedIn uses Big Data to connect job seekers with job opportunities. With LinkedIn, headhunters no longer need to cold call potential employees. They can find and contact them via a simple search. Similarly, job seekers can get a warm introduction to a potential hiring manager by connecting to others on the site.

LinkedIn CEO Jeff Weiner recently talked about the future of the site and its economic graph—a digital map of the global economy that will in real time identify "the trends pointing to economic opportunities."[2] The challenge of delivering on such a graph and its predictive capabilities is a Big Data problem.

---

[2]http://www.linkedin.com/today/post/article/20121210053039-22330283-the-future-of-linkedin-and-the-economic-graph

Second, both of these companies went public in just the last few years—Facebook on NASDAQ, LinkedIn on NYSE. Although these companies and Google are consumer companies on the surface, they are really massive Big Data companies at the core.

The public offerings of these companies—combined with that of Splunk, a provider of operational intelligence software, and that of Tableau Software, a visualization company—significantly increased Wall Street's interest in Big Data businesses.

As a result, venture capitalists in Silicon Valley are lining up to fund Big Data companies like never before. Big Data is defining the next major wave of startups that Silicon Valley is hoping to take to Wall Street over the next few years.

Accel Partners, an early investor in Facebook, announced a $100 million Big Data Fund in late 2011 and made its first investment from the fund in early 2012. Zetta Venture Partners is a new fund launched in 2013 focused exclusively on Big Data analytics. Zetta was founded by Mark Gorenberg, who was previously a Managing Director at Hummer Winblad.[3] Well-known investors Andreessen Horowitz, Greylock Partners, and others have made a number of investments in the space as well.

Third, business people, who are active users of Amazon, Facebook, LinkedIn, and other consumer products with data at their core, started expecting the same kind of fast and easy access to Big Data at work that they were getting at home. If Internet retailer Amazon could use Big Data to recommend books to read, movies to watch, and products to purchase, business users felt their own companies should be able to leverage Big Data too.

Why couldn't a car rental company, for example, be smarter about which car to offer a renter? After all, the company has information about which car the person rented in the past and the current inventory of available cars. But with new technologies, the company also has access to public information about what's going on in a particular market—information about conferences, events, and other activities that might impact market demand and availability.

By bringing together internal supply chain data with external market data, the company should be able to predict which cars to make available and when more accurately.

Similarly, retailers should be able to use a mix of internal and external data to set product prices, placement, and assortment on a day-to-day basis. By taking into account a variety of factors—from product availability to consumer shopping habits, including which products tend to sell well together—retailers

---

[3]Zetta Venture Partners is an investor in my company, Content Analytics.

can increase average basket size and drive higher profits. This in turn keeps their customers happy by having the right products in stock at the right time.

So while Big Data became hot seemingly overnight, in reality, Big Data is the culmination of a mix of years of software development, market growth, and pent up consumer and business user demand.

How Google Puts Big Data Initiatives to Work

If there's one technology company that has capitalized on that demand and that epitomizes Big Data, it's search engine giant Google, Inc. According to Google, the company handles an incredible 100 billion search queries per month.[4]

But Google doesn't just store links to the web sites that appear in its search results. It also stores all the searches people make, giving the company unparalleled insight into the when, what, and how of human search behavior.

Those insights mean that Google can optimize the advertising it displays to monetize web traffic better than almost every other company on the planet. It also means that Google can predict what people are going to search for next. Put another way, Google knows what you're looking for before you do!

Google has had to deal, for years, with massive quantities of unstructured data such as web pages, images, and the like rather than more traditional structured data, such as tables that contain names and addresses. As a result, Google's engineers developed innovative Big Data technologies from the ground up. Such opportunities have helped Google attract an army of talented engineers who are attracted to the unique size and scale of Google's technical challenges.

Another advantage the company has is its infrastructure. The Google search engine itself is designed to work seamlessly across hundreds of thousands of servers. If more processing or storage is required or if a server goes down, Google's engineers simply add more servers. Some estimates put Google's total number of servers at greater than a million.

Google's software technologies were designed with this infrastructure in mind. Two technologies in particular, MapReduce and the Google File System, "reinvented the way Google built its search index," *Wired* magazine reported during the summer of 2012.[5]

---

[4]http://phandroid.com/2014/04/22/100-billion-google-searches/
[5]http://www.wired.com/wiredenterprise/2012/08/googles-mind-blowing-big-data-tool-grows-open-source-twin/

Numerous companies are now embracing Hadoop, an open-source derivative of MapReduce and the Google File System. Hadoop, which was pioneered at Yahoo! based on a Google paper about MapReduce, allows for distributed processing of large data sets across many computers.

While other companies are just now starting to make use of Hadoop, Google has been using large-scale Big Data technologies for years, giving it an enormous leg up in the industry. Meanwhile, Google is shifting its focus to other, newer technologies. These include Caffeine for content indexing, Pregel for mapping relationships, and Dremel for querying very large quantities of data. Dremel is the basis for the company's BigQuery offering.[6]

Now Google is opening up some of its investment in data processing to third parties. Google BigQuery is a web offering that allows interactive analysis of massive data sets containing billions of rows of data. BigQuery is data analytics on-demand, in the cloud. In 2014, Google introduced Cloud Dataflow, a successor to Hadoop and MapReduce, which works with large volumes of both batch-based and streaming-based data.

Previously, companies had to buy expensive installed software and set up their own infrastructure to perform this kind of analysis. With offerings like BigQuery, these same companies can now analyze large data sets without making a huge up-front investment.

Google also has access to a very large volume of machine data generated by people doing searches on its site and across its network. Every time someone enters a search query, Google knows what that person is looking for. Every human action on the Internet leaves a trail, and Google is well positioned to capture and analyze that trail.

Yet Google has even more data available to it beyond search. Companies install products like Google Analytics to track visitors to their own web sites, and Google gets access to that data too. Web sites use Google AdSense to display ads from Google's network of advertisers on their own web sites, so Google gets insight not only into how advertisements perform on its own site but on other publishers' sites as well. Google also has vast amounts of mapping data from Google Maps and Google Earth.

Put all that data together and the result is a business that benefits not just from the best technology but from the best information. When it comes to Information Technology (IT), many companies invest heavily in the *technology* part of IT, but few invest as heavily and as successfully as Google does in the *information* component of IT.

---

[6]http://www.wired.com/wiredenterprise/2012/08/googles-dremel-makes-big-data-look-small/

■ **Note** When it comes to IT, the most forward thinking companies invest as much in information as they do in technology.

How Big Data Powers Amazon's Quest to Become the World's Largest Retailer

Of course, Google isn't the only major technology company putting Big Data to work. Internet retailer Amazon.com has made some aggressive moves and may pose the biggest long-term threat to Google's data-driven dominance.

At least one analyst predicts that Amazon will exceed $100B in revenue by 2015, putting it on track to eclipse Walmart as the world's largest retailer. Like Google, Amazon has vast amounts of data at its disposal, albeit with a much heavier e-commerce bent.

Every time a customer searches for a TV show to watch or a product to buy on the company's web site, Amazon gets a little more insight about that customer. Based on searches and product purchasing behavior, Amazon can figure out what products to recommend next.

And the company is even smarter than that. It constantly tests new design approaches on its web site to see which approach produces the highest conversion rate.

Think a piece of text on a web page on the Amazon site just happened to be placed there? Think again. Layout, font size, color, buttons, and other elements of the company's site design are all meticulously tested and retested to deliver the best results.

The data-driven approach doesn't stop there. According to more than one former employee, the company culture is ruthlessly data-driven. The data shows what's working and what isn't, and cases for new business investments must be supported by data.

This incessant focus on data has allowed Amazon to deliver lower prices and better service. Consumers often go directly to Amazon's web site to search for goods to buy or to make a purchase, skipping search engines like Google entirely.

The battle for control of the consumer reaches even further. Apple, Amazon, Google, and Microsoft—known collectively as The Big Four—are battling it out not just online but in the mobile domain as well.

With consumers spending more and more time on mobile phones and tablets instead of in front of their computers, the company whose mobile device is in the consumer's hand will have the greatest ability to sell to that consumer and gain the most insight about that consumer's behavior. The more information a company has about consumers in aggregate and as individuals, the more effectively it can target its content, advertisements, and products to those consumers.

Incredibly, Amazon's grip reaches all the way from the infrastructure supporting emerging technology companies to the mobile devices on which people consume content. Years ago, Amazon foresaw the value in opening the server and storage infrastructure that is the backbone of its e-commerce platform to others.

Amazon Web Services (AWS), as the company's public cloud offering is known, provides scalable computing and storage resources to emerging and established companies. While AWS is still relatively early in its growth, one analyst estimate puts the offering at greater than a $3.8 billion annual revenue run rate.[7]

The availability of such easy-to-access computing power is paving the way for new Big Data initiatives. Companies can and will still invest in building out their own private infrastructure in the form of private clouds, of course. Private clouds—clouds that companies manage and host internally—make sense when dealing with specific security, regulatory, or availability concerns.

But if companies want to take advantage of additional or scalable computing resources quickly, they can simply fire up a bunch of server instances in Amazon's public cloud. What's more, Amazon continues to lower the prices of its computing and storage offerings. Because of the company's massive purchasing power and the scale of its infrastructure, it can negotiate prices for computers and networking equipment that are far lower than those available even to most other large corporations. Amazon's Web Services offering puts the company front and center not just with its own consumer-facing site and mobile devices like the Kindle Fire, but with infrastructure that supports thousands of other popular web sites as well.

The result is that Big Data analytics no longer requires investing in fixed-cost IT up-front. Users can simply purchase more computing power to perform analysis or more storage to store their data when they need it. Data capture and analysis can be done quickly and easily in the cloud, and users don't need to make expensive decisions about IT infrastructure up-front. Instead they can purchase just the computing and storage resources they need to meet their Big Data needs and do so at the time and for the duration that those resources are actually needed.

---

[7]http://www.zdnet.com/amazons-aws-3-8-billion-revenue-in-2013-says-analyst-7000009461/

Businesses can now capture and analyze an unprecedented amount of data—data they simply couldn't afford to analyze or store before and instead had to throw away.

Note   One of the most powerful aspects of Big Data is its scalability. Using cloud resources, including analytics and storage, there is now no limit to the amount of data a company can store, crunch, and make useful.

Big Data Finally Delivers the Information Advantage

Infrastructure like Amazon Web Services combined with the availability of open-source technologies like Hadoop means that companies are finally able to realize the benefits long promised by IT.

For decades, the focus in IT was on the T—the technology. The job of the Chief Information Officer (CIO) was to buy and manage servers, storage, and networks.

Now, however, it is information and the ability to store, analyze, and predict based on that information that is delivering a competitive advantage (Figure 1-1).

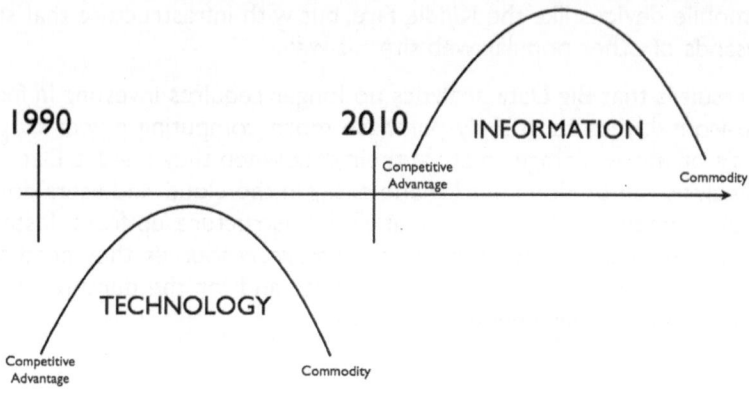

Figure 1-1. Information is becoming the critical asset that technology once was

When IT first became widely available, companies that adopted it early on were able to move faster and out-execute those that did not. Some credit Microsoft's rise in the 1990s not just to its ability to deliver the world's most widely used operating system, but to the company's internal embrace of email as the standard communication mechanism.

While many companies were still deciding whether or how to adopt email, at Microsoft, email became the de facto communication mechanism for discussing new hires, product decisions, marketing strategy, and the like. While electronic group communication is now commonplace, at the time it gave the company a speed and collaboration advantage over those companies that had not yet embraced email.

Companies that embrace data and democratize the use of that data across their organizations will benefit from a similar advantage. Companies like Google and Facebook have already benefited from this data democratization.

By opening up their internal data analytics platforms to analysts, managers, and executives throughout their organizations, Google, Facebook, and others have enabled everyone in their organizations to ask business questions of the data and get the answers they need, and to do so quickly. As Ashish Thusoo, a former Big Data leader at Facebook, put it, new technologies have changed the conversation from "what data to store" to "what can we do with more data?"

Facebook, for example, runs its Big Data effort as an internal service. That means the service is designed not for engineers but for end-users—line managers who need to run queries to figure out what's working and what isn't.

As a result, managers don't have to wait days or weeks to find out what site changes are most effective or which advertising approaches work best. They can use the internal Big Data service to get answers to their business questions in real time. And the service is designed with end-user needs in mind, all the way from operational stability to social features that make the results of data analysis easy to share with fellow employees.

The past two decades were about the technology part of IT. In contrast, the next two decades will be about the information part of IT. Companies that can process data faster and integrate public and internal sources of data will gain unique insights that enable them to leapfrog over their competitors.

As J. Andrew Rogers, founder and CTO of the Big Data startup SpaceCurve, put it, "the faster you analyze your data, the greater its predictive value." Companies are moving away from batch processing (that is, storing data and then running slow analytics processing on the data after the fact) to real-time analytics to gain a competitive advantage.

The good news for executives is that the information advantage that comes from Big Data is no longer exclusively available to companies like Google and Amazon. Open-source technologies like Hadoop are making it possible for many other companies—both established *Fortune* 1,000 enterprises and emerging startups—to take advantage of Big Data to gain a competitive advantage, and to do so at a reasonable cost. Big Data truly does deliver the long-promised information advantage.

What Big Data Is Disrupting

The big disruption from Big Data is not just the ability to capture and analyze more data than in the past, but to do so at price points that are an order of magnitude cheaper. As prices come down, consumption goes up.

This ironic twist is known as *Jevons paradox*, named for the economist who made this observation about the Industrial Revolution. As technological advances make storing and analyzing data more efficient, companies are doing a lot more analysis, not less. This, in a nutshell, is what's so disruptive about Big Data.

Many large technology companies, from Amazon to Google and from IBM to Microsoft, are getting in on Big Data. Yet dozens of startups are cropping up to deliver open-source and cloud-based Big Data solutions.

While the big companies are focused on horizontal Big Data solutions—platforms for general-purpose analysis—smaller companies are focused on delivering applications for specific lines of business and key verticals. Some products optimize sales efficiency while others provide recommendations for future marketing campaigns by correlating marketing performance across a number of different channels with actual product usage data. There are Big Data products that can help companies hire more efficiently and retain those employees once hired.

Still other products analyze massive quantities of survey data to provide insights into customer needs. Big Data products can evaluate medical records to help doctors and drug makers deliver better medical care. And innovative applications can now use statistics from student attendance and test scores to help students learn more effectively and have a higher likelihood of completing their studies.

Historically, it has been all too easy to say that we don't have the data we need or that the data is too hard to analyze. Now, the availability of these Big Data Applications means that companies don't need to develop or deploy all Big Data technology in-house. In many cases they can take advantage of cloud-based services to address their analytics needs. Big Data is making data, and the ability to analyze that data and gain actionable insights from it, much, much easier than it has been. That truly is disruptive.

Big Data Applications Changing Your Work Day

Big Data Applications, or BDAs, represent the next big wave in the Big Data space. Industry analyst firm CB Insights looked at the funding landscape for Big Data and reported that Big Data companies raised some $1.28 billion in the first half of 2013 alone.[8] Since then, investors have continued to pour money into existing infrastructure players. One company, Cloudera, a commercial provider of Hadoop software, announced a massive $900 million funding round in March of 2014, bringing the company's total funding to $1.2 billion.

Going forward, the focus will shift from the infrastructure necessary to work with large amounts of data to the uses of that data. No longer will the question be where and how to store large quantities of data. Instead, users will ask how they can use all that data to gain insight and obtain competitive advantage.

Note The era of creating and providing the infrastructure necessary to work with Big Data is nearly over. Going forward, the focus will be on one key question: "How can we use all our data to create new products, sell more, and generally outrun our competitors?"

Splunk, an operational intelligence company, is one existing example of this. Historically, companies had to analyze *log files*—the files generated by network equipment and servers that make up their IT systems—in a relatively manual process using scripts they developed themselves.

Not only did IT administrators have to maintain the servers, network equipment, and software for the infrastructure of a business, they also had to build their own tools in the form of scripts to determine the cause of issues arising from those systems. And those systems generate an immense amount of data. Every time a user logs in or a file is accessed, every time a piece of software generates a warning or an error, that is another piece of data that administrators have to comb through to figure out what's going on.

With BDAs, companies no longer have to build the tools themselves. They can take advantage of pre-built applications and focus on running their businesses instead. Splunk's software, for example, makes it possible to find infrastructure issues easily by searching through IT log files and visualizing the locations and frequency of issues. Of course, the company's software is *primarily installed software,* meaning it has to be installed at a customer's site.

---

[8]http://www.cbinsights.com/blog/big-data-funding-venture-capital-2013

Cloud-based BDAs hold the promise of not requiring companies to install any hardware or software at all. In some ways, they can be thought of as the next logical step after Software as a Service (SaaS) offerings. SaaS, which are software products delivered over the Internet, are relatively well-established. As an example, Salesforce.com, which first introduced the "no software" concept over a decade ago, has become the de-facto standard for cloud-based Customer Relationship Management (CRM), software that helps companies manage their customer lists and relationships.

SaaS transformed software into something that could be used anytime, anywhere, with little maintenance required on the part of its users. Just as SaaS transformed how we access software, BDAs are transforming how we access data. Moreover, BDAs are moving the value in software from the software itself to the data that that software enables us to act on. Put another way, BDAs have the potential to turn today's technology companies into tomorrow's highly valuable information businesses.

BDAs are transforming both our workdays and our personal lives, often at the same time. Opower, for example, is changing the way energy is consumed. The company tracks energy consumption across some 50 million U.S. households by working with 75 different utility companies. The company uses data from smart meters—devices that track household energy usage—to provide consumers with detailed reports on energy consumption. Even a small change in energy consumption can have a big impact when spread across tens of millions of households.

Just as Google has access to incredible amounts of data about how consumers behave on the Internet, Opower has huge amounts of data about how people behave when it comes to energy usage. That kind of data will ultimately give Opower, and companies like it, highly differentiated insights. Although the company has started out by delivering energy reports, by continuing to build up its information assets, it will be well-positioned as a Big Data business.[9]

BDAs aren't just appearing in the business world, however. Companies are developing many other data applications that can have a positive impact on our daily lives. In one example, some mobile applications track health-related metrics and make recommendations to improve human behavior. Such products hold the promise of reducing obesity, increasing quality of life, and lowering healthcare costs. They also demonstrate how it is at the intersection of new mobile devices, Big Data, and cloud computing where some of the most innovative and transformative Big Data Applications may yet appear.

---

[9]http://www.forbes.com/sites/davefeinleib/2012/10/24/software-is-dead-long-live-big-data-2/

Big Data Enables the Move to Real Time

If the last few years of Big Data have been about capturing, storing, and analyzing data at lower cost, the next few years will be about speeding up the access to that data and enabling us to act on it in real time. If you've ever clicked on web site button only to be presented with a wait screen, you know just how frustrating it is to have to wait for a transaction to complete or for a report to be generated.

Contrast that with the response time for a Google search result. Google Instant, which Google introduced in 2010, shows you search results as you type. By introducing the feature, Google ended up serving five to seven times more search result pages for typical searches. When the interface was introduced, people weren't[10] sure they liked it. Now, just a few years later, no one can imagine living without it.

Data analysts, managers, and executives want the Google Instant kind of immediacy in understanding their businesses. As these users of Big Data push for faster and faster results, just adopting Big Data technologies will no longer be sufficient. Sustained competitive advantage will come not from Big Data itself but from the ability to gain insight from information assets faster than others. Interfaces like Google Instant demonstrate just how powerful immediate access can be.

According to IBM, "every day we create 2.5 quintillion bytes of data—so much that 90% of the data in the world has been created in the last two years alone."[11] Industry research firm Forrester estimates that the overall amount of corporate data is growing by 94% per year.[12]

With this kind of growth, every company needs a Big Data roadmap. At a minimum, companies need to have a strategy for capturing data, from machine log files generated by in-house computer systems to user interactions on web sites, even if they don't decide what to do with that data until later. As Rogers put it, "data has value far beyond what you originally anticipate—don't throw it away."

■ **Tip**   "Data has value far beyond what you originally anticipate—don't throw it away." —J. Andrew Rogers, CTO, SpaceCurve.

---

[10]http://googleblog.blogspot.com/2010/09/google-instant-behind-scenes.html
[11]http://www-01.ibm.com/software/data/bigdata/what-is-big-data.html
[12]http://www.forbes.com/sites/ciocentral/2012/07/05/best-practices-for-managing-big-data/

Companies need to plan for exponential growth of their data. While the number of photos, instant messages, and emails is very large, the amount of data generated by networked "sensors" such as mobile phones, GPSs, and other devices is much larger.

Ideally, companies should have a vision for enabling data analysis throughout the organization and for that analysis to be done in as close to real time as possible. By studying the Big Data approaches of Google, Amazon, Facebook, and other tech leaders, you can see what's possible with Big Data. From there, you can put an effective Big Data strategy in place in your own organization.

Companies that have success with Big Data add one more key element to the mix: a Big Data leader. All the data in the world means nothing if you can't get insights from it. Your Big Data leader—a Chief Data Officer or a VP of Data Insights—can not only help your entire organization get the right strategy in place but can also guide your organization in getting the actionable insights it needs.

Companies like Google and Amazon have been using data to drive their decisions for years and have become wildly successful in the process. With Big Data, these same capabilities are now available to you. You'll read a lot more about how to take advantage of these capabilities and formulate your own Big Data roadmap in Chapter 3. But first, let's take a look at the incredible market drivers and innovative technologies all across the Big Data landscape.[13]

---

[13]Some of the material for this chapter appeared in a guest contribution I authored for *Harvard Business Review China*, January 2013.

The Big Data Landscape

Infrastructure and Applications

Now that we've explored a few aspects of Big Data, we'll take a look at the broader landscape of companies that are playing a role in the Big Data ecosystem. It's easiest to think about the Big Data landscape in terms of infrastructure and applications.

The chart that follows (see Figure 2-1) is "The Big Data Landscape." The landscape categorizes many of the players in the Big Data space. Since new entrants emerge regularly, the latest version of the landscape is always available on the web at www.bigdatalandscape.com.

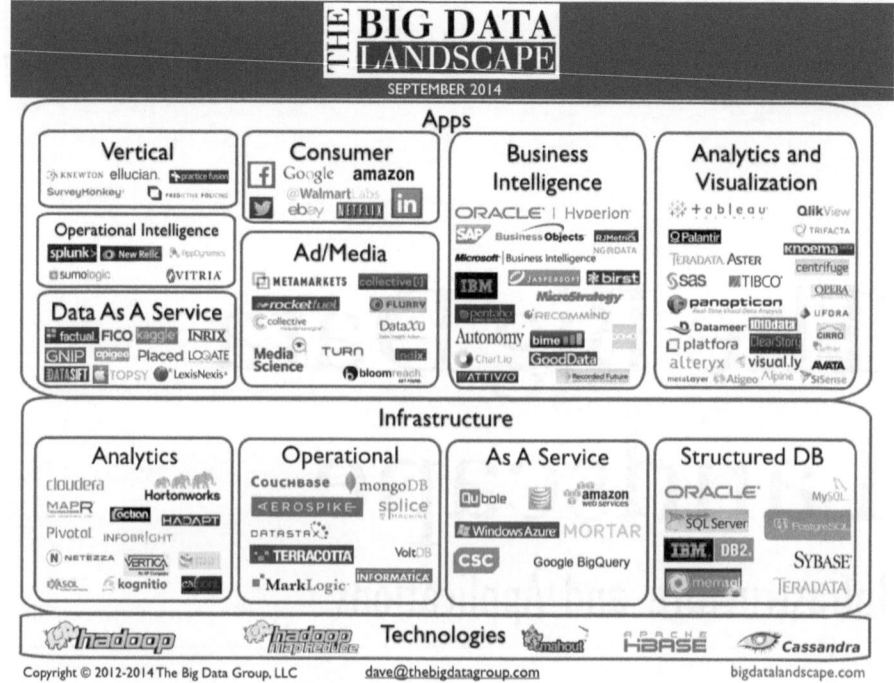

Figure 2-1. The Big Data landscape

Infrastructure is primarily responsible for storing and to some extent processing the immense amounts of data that companies are capturing. Humans and computer systems use applications to gain insights from data.

People use applications to visualize data so they can make better decisions, while computer systems use applications to serve up the right ads to the right people or to detect credit card fraud, among many other activities. Although we can't touch on every company in the landscape, we will describe a number of them and how the ecosystem came to be.

Big Data Market Growth

Big Data is a big market. Market research firm IDC expects the Big Data market to grow to $23.8 billion a year by 2016 and that the growth rate in the space will be 31.7% annually. That doesn't[1] even include analytics software, which by itself counts for another $51 billion.

---

[1] http://gigaom.com/2013/01/08/idc-says-big-data-will-be-24b-market-in-2016-i-say-its-bigger/

The amount of data we're generating is growing at an astounding rate. One of the most interesting measures of this is Facebook's growth. In October 2012, the company announced it had hit one billion users—nearly 15% of the world's population. Facebook had more than 1.23 billion users worldwide by the end of 2013 (see Figure 2-2), adding 170 million users in the year. The company has had to develop a variety of new infrastructure and analytics technologies to keep up with its immense user growth.[2]

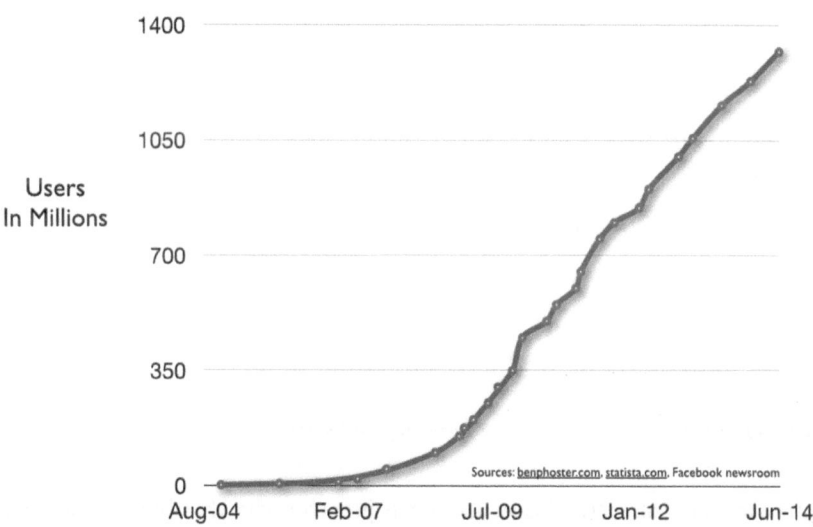

Figure 2-2. Facebook's user growth rate

Facebook handles some 350 million photo uploads, 4.5 billion Likes, and 10 billion messages every day. That means the company stores more than 100 petabytes when it comes to the data it uses for analytics and ingests more than 500 terabytes of new data per day.[3,4] That's the equivalent of adding the data stored on roughly 2,000 Macintosh Air hard drives if they were all fully used.

---

[2]http://www.theguardian.com/technology/2014/feb/04/facebook-10-years-mark-zuckerberg
[3]http://www.digitaltrends.com/social-media/according-to-facebook-there-are-350-million-photos-uploaded-on-the-social-network-daily-and-thats-just-crazy/#!Cht6e
[4]http://us.gizmodo.com/5937143/what-facebook-deals-with-everyday-27-billion-likes-300-million-photos-uploaded-and-500-terabytes-of-data

Twitter provides another interesting measure of data growth. The company reached more than 883 million registered users as of 2013 and is handling more than 500 million tweets per day, up from 20,000 per day just five years earlier (see Figure 2-3).[5]

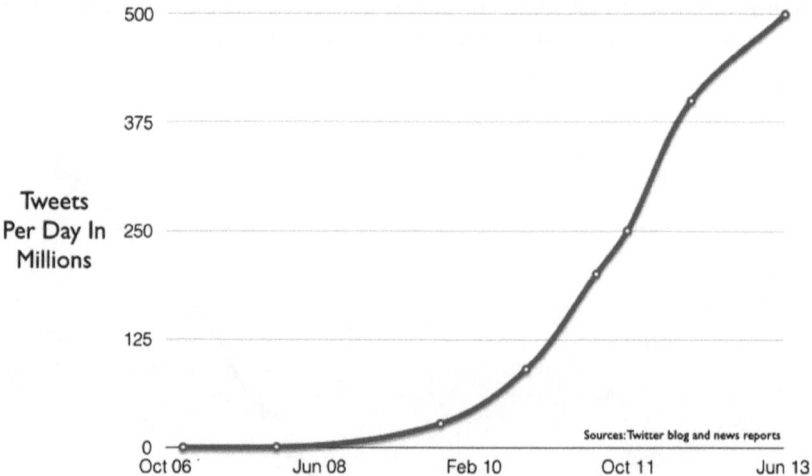

Figure 2-3. *Growth of tweets on social networking site Twitter*

To put this in perspective, however, this is just the data that human beings generate. Machines are generating even more data. Every time we click on a web site button, make a purchase, call someone on the phone, or do virtually any other activity, we leave a digital trail. The simple action of uploading a photo generates lots of other data: who uploaded the photo and when, who it was shared with, what tags are associated with it, and so on.

The volume of data is growing all around us: Walmart handles more than a million customer transactions every hour, and about 90 trillion emails are sent every year. Ironically, more than half of all that email, 71.8% of it, is considered spam. And the volume of business data doubles every 1.2 years, according to one estimate.[6,7]

To address this incredible growth, a number of new companies have emerged and a number of existing companies are repositioning themselves and their offerings around Big Data.

[5]http://twopcharts.com/twitteractivitymonitor
[6]http://www.securelist.com/en/analysis/204792243/Spam_in_July_2012
[7]http://knowwpcarey.com/article.cfm?cid=25&aid=1171

The Role of Open Source

Open source has played a significant role in the recent evolution of Big Data. But before we talk about that, it's important to give some context on the role of open source more generally.

Just a few years ago, Linux became a mainstream operating system and, in combination with commodity hardware (low cost, off-the-shelf servers), it cannibalized vendors like Sun Microsystems that were once dominant. Sun, for example, was well known for its version of Unix, called Solaris, which ran on its custom SPARC hardware.

With Linux, enterprises were able to use an open source operating system on low-cost hardware to get much of the same functionality, at a much lower cost. The availability of open source database MySQL, open source web server Apache, and open source scripting language PHP, which was originally created for building web sites, also drove the popularity of Linux.

As enterprises began to use and adopt Linux for large-scale commercial use, they required enterprise-grade support and reliability. It was fine for engineers to work with open source Linux in the lab, but businesses needed a vendor they could call on for training, support, and customization. Put another way, big companies like buying from other big companies.

Among a number of vendors, Red Hat emerged as the market leader in delivering commercial support and service for Linux. The company now has a market cap of just over $10 billion. MySQL AB, a Swedish company, sponsored the development of the open source MySQL database project. Sun Microsystems acquired MySQL AB for $1 billion in early 2008 and Oracle acquired Sun in late 2009.

Both IBM and Oracle, among others, commercialized large-scale relational databases. Relational databases allow data to be stored in well-defined tables and accessed by a key. For example, an employee might be identified by an employee number, and that number would then be associated with a number of other fields containing information about the employee, such as her name, address, hire date, and position.

Such databases worked well until companies had to contend with really large quantities of unstructured data. Google had to deal with huge numbers of web pages and the relationships between links in those pages. Facebook had to contend with social graph data. The social graph is the digital representation of the relationships between people on its social network and all of the unstructured data at the end of each point in the graph, such as photos, messages, and profiles. These companies also wanted to take advantage of the lower costs of commodity hardware.

■ **Note**   Relational databases, the workhorses of many a large corporate IT environments, worked beautifully until companies realized they were generating far more unstructured data and needed a different solution. Enter NoSQL and graph databases, among others, which are designed for today's diverse data environments.

So companies like Google, Yahoo!, Facebook, and others developed their own solutions for storing and processing vast quantities of data. In the operating system and database markets, Linux emerged as an open source version of Unix and MySQL as an open source alternative to databases like Oracle. Today, much the same thing is happening in the Big Data world.

Apache Hadoop, an open source distributed computing platform for storing large quantities of data via the Hadoop Distributed File System (HDFS)—and dividing operations on that data into small fragments via a programming model called MapReduce—was derived from technologies originally built at Google and Yahoo!.

Other open source technologies have emerged around Hadoop. Apache Hive provides data warehousing capabilities, including data extract/transform/load (ETL), a process for extracting data from a variety of sources, transforming it to fit operational needs (including ensuring the quality of the data), and loading it into the target database. Apache HBase provides real-time read-write access to very large structured tables on top of Hadoop. It is modeled on Google's[8] BigTable. Meanwhile, Apache Cassandra provides fault-tolerant data storage by replicating data.

Historically, such capabilities were only available from commercial software vendors, typically on specialized hardware. Linux made the capabilities of Unix available on commodity hardware, drastically reducing the cost of computing. In much the same way, open source Big Data technologies are making data storage and processing capabilities that were previously only available to companies like Google or from commercial vendors available to everyone on commodity hardware.

The widespread availability of low-cost Big Data technology reduces the up-front cost of working with Big Data and has the potential to make Big Data accessible to a much larger number of potential users. Closed-source vendors point out that while open source software is free to adopt, it can be costly to maintain, especially at scale.

---

[8]http://en.wikipedia.org/wiki/Extract,_transform,_load

That said, the fact that open source is free to get started with has made it an appealing option for many. Some commercial vendors have adopted "freemium" business models to compete. Products are free to use on a personal basis or for a limited amount of data, but customers are required to pay for departmental or larger data usage.

Those enterprises that adopt open source technologies over time require commercial support for them, much as they did with Linux. Companies like Cloudera, Hortonworks, and MapR are addressing that need for Hadoop, while companies like DataStax are doing the same for Cassandra. By the same token, LucidWorks is performing such a role for Apache Lucerne, an open source text search engine used for indexing and searching large quantities of web pages and documents.

Note   Even though companies like Cloudera and Hortonworks provide products built on open source software (like Hadoop, Hive, Pig, and others), companies will always need commercially packaged, tested versions along with support and training. Hence Intel's recent infusion of $900 million into Cloudera.

Enter the Cloud

Two other market trends are occurring in parallel. First, the volume of data is increasing—doubling almost every year. We are generating more data in the form of photos, tweets, likes, and emails. Our data has data associated with it. What's more, machines are generating data in the form of status updates and other information from servers, cars, airplanes, mobile phones, and other devices.

As a result, the complexity of working with all that data is increasing. More data means more data to integrate, understand, and try to get insights from. It also means higher risks around data security and data privacy. And while companies historically viewed internal data such as sales figures and external data like brand sentiment or market research numbers separately, they now want to integrate those kinds of data to take advantage of the resulting insights.

Second, enterprises are moving computing and processing to the cloud. This means that instead of buying hardware and software and installing it in their own data centers and then maintaining that infrastructure, they're now getting the capabilities they want on demand over the Internet. As mentioned, Software as a Service (SaaS) company Salesforce.com pioneered the delivery of applications over the web with its "no software" model for customer relationship management (CRM). The company has continued to build out an ecosystem of offerings to complement its core CRM solution. The SaaS model has gained momentum with a number of other companies offering cloud-based services targeted at business users.

Meanwhile, Amazon paved the way for infrastructure on demand with Amazon Web Services (AWS), the company's extremely popular cloud-based computing and storage offering. Amazon launched AWS in 2002 with the idea that it could make a profit on the infrastructure required to run the Amazon.com store. The company has continued to add on-demand infrastructure services that allow developers to bring up new servers, storage, and databases quickly.[9,10]

Amazon has also introduced Big Data-specific services, including Amazon Elastic MapReduce (EMR), an Amazon cloud-based version of the open source Hadoop MapReduce offering. Amazon Redshift is a data-warehousing on-demand solution that Amazon expects will cost as little as $1,000 per terabyte per year, less than a tenth of what companies typically pay for on-premise data-warehousing, which can run to more than $20,000 per terabyte annually. Meanwhile, Amazon Glacier provides low-cost digital archiving services at $0.01 per gigabyte per month or about $120 per terabyte per year.[11]

Amazon has two primary advantages over other providers. It has a very well-known consumer brand. The company also benefits from the economies of scale it gets from supporting the Amazon.com web site as well as from serving so many customers on its infrastructure. Other technology market leaders also offer cloud infrastructure services, including Google with its Google Cloud Platform and Microsoft with Windows Azure, but Amazon has paved the way and grabbed pole position with AWS.

AWS has seen incredible growth. The service was expected to bring in $3.2 billion in revenue for the company in 2013. As of June, 2012, Amazon was storing more than a trillion objects in its Simple Storage Service (S3) and was adding more than 40,000 new objects per second. That number is up from just 2.9 billion objects stored as of the end of 2006 and 262 billion at the end of 2010. Major companies like Netflix, Dropbox, and others run their business on AWS.[12,13]

Amazon continues to broaden its infrastructure-on-demand offerings, adding services for IP routing, email sending, and a host of Big Data related services. In addition to offering its own infrastructure on demand, Amazon also works with an ecosystem of partners to offer their infrastructure products. Thus, the headline for any new infrastructure startup thinking about building a public cloud offering may well be the following: find a way to partner with Amazon or expect the company to come out with a competitive offering.

---

[9]http://phx.corporate-ir.net/phoenix.zhtml?c=176060&p=irol-corporateTimeline
[10]http://en.wikipedia.org/wiki/Amazon_Web_Services#cite_note-1
[11]http://www.informationweek.com/software/information-management/amazon-redshift-leaves-on-premises-openi/240143912
[12]http://www.informationweek.com/cloud/infrastructure-as-a-service/amazon-web-services-revenue-new-details/d/d-id/1112068?
[13]http://aws.typepad.com/aws/2012/06/amazon-s3-the-first-trillion-objects.html

With cloud services, customers pay only for what they use. This makes it significantly easier to get new Big Data projects up and running. Companies are no longer constrained by the often expensive up-front costs typically associated with buying, deploying, and managing server and storage infrastructure.

We can store and analyze the massive quantities of data associated with Big Data initiatives much more quickly and cost effectively than in the past. With cloud-based Big Data infrastructure, it is possible to go from planning to insights in days instead of months or years. This in turn means more innovation at both startups and big companies in the area of Big Data as well as far easier access to data-driven insights.

■ **Note** Affordable cloud-based Big Data infrastructure means companies can go from design to product, or from hypothesizing to discovering insights, in days instead of months or years.

Overcoming the Challenges of the Cloud

Of course, many are still skeptical of taking advantage of public cloud infrastructure. Historically, there have been three major potential issues with public cloud services. First, technologists at large enterprises felt that such services were not secure and that in-house infrastructure was more secure. Second, many large vendors simply didn't offer Internet/cloud-based versions of their software. Companies had to buy the hardware and run the software themselves or hire a third party to do so. Finally, it was difficult to get large volumes of data from in-house systems into the cloud.

While the first challenge remains true for certain government agencies and enterprises subject to compliance requirements, the widespread adoption of services like Salesforce for managing customer data and Box, Dropbox, and Google Drive for storing files illustrates that many companies are now comfortable having third-party vendors store their confidential information. Simply put, businesses are adopting far more applications delivered over the net than they have in the past.

First it was customer data, then HR data. Today, the market values Workday, a web-based provider of HR management solutions, at nearly $9 billion. Over time, enterprises will migrate more and more of their business applications and data to the cloud.

The third challenge, moving massive amounts of data to the cloud, remains an issue. Many experts feel that when it comes to really high volume data, data is subject to a sort of gravitational pull. They believe that data that starts on-premise, at a company, will remain there, while data that starts in the cloud will stay there.

But as more line-of-business applications become available over the net, more data will start in the cloud and will stay there. In addition, companies are emerging to provide technologies that speed up the transfer of large quantities of data. Aspera, for example, has focused on accelerating file transfer speeds, particularly for large audio and video files. Companies like Netflix and Amazon use Aspera to transfer files at up to 10 times traditional speeds.

With the cloud, companies get a host of other advantages: they spend less time maintaining and deploying hardware and software, and they can scale on-demand. If a company needs more computing resources or storage, that's not a matter of months but of minutes. What's more, enterprises with installed software have traditionally lagged behind in terms of using the latest version of the software. With applications that come over the net, users can be on the latest version as soon as it's available.

Of course, there are tradeoffs. Companies are at the mercy of the public cloud provider they choose to work with regarding costs. But competition between cloud vendors has continued to push prices down. Once a company has built and deployed applications on top of a particular cloud vendor, switching platforms can be challenging. Although the underlying infrastructures are similar, scripts for deploying new application builds, hooks into platform-specific services, and network configurations are all based on the chosen platform.

■ **Note** Most companies tend to stick with the public cloud platform they initially choose. That's one reason the battle for cloud customers is more intense than ever.

Customers also depend on such providers to deliver reliable service. Amazon has suffered a few major high-profile outages that have caused some to question whether relying on its service makes sense. One such outage caused movie-streaming provider Netflix to lose service on Christmas Eve and Christmas Day 2012, traditionally a very popular time for watching movies. Yet in general, cloud-based infrastructure services have continued to become more transparent about their service availability and have added enterprise-grade service and support options to meet the requirements of business customers. In many cases, providers like Amazon, Google, and Microsoft may deliver infrastructure that is more reliable, secure, and cost-effective than many enterprises can deliver themselves, due to the incredible scale at which these providers operate.

Cloud providers will continue to introduce more infrastructure-on-demand capabilities. As they compete with each other and reduce prices, even more businesses will take advantage of such offerings. Startups will introduce new, innovative Big Data applications built on top of these infrastructure services and enterprises and consumers will both benefit.

Big Data Infrastructure

With the context of open source and Big Data in the cloud in mind, we'll now take a look at some of the companies playing key roles in the infrastructure and applications spaces.

In the larger context, investors had until recently put a relatively small amount of capital to work in Big Data infrastructure. Their investments in Hadoop-related companies Cloudera, Hortonworks, and MapR, and in NoSQL companies MongoDB, Couchbase, and others, totaled less than a billion through the end of 2013. In early 2014, however, Intel and Cloudera made headlines when Intel invested a whopping $900 million into Cloudera.

Cloudera has been the most visible of all of the new Big Data infrastructure companies. Cloudera sells tools and consulting services that help companies run Hadoop. Its investors include Accel Partners, Ignition Partners, In-Q-Tel, Intel, Greylock Partners, and Meritech Capital Partners. Cloudera was founded by the marquee team of Mike Olson, Amr Awadallah, who worked with Hadoop at Yahoo!, Jeff Hammerbacher, who had worked with it at Facebook, and Christophe Bisciglia from Google.[14,15]

It is interesting to note that Google first published a paper describing Google MapReduce and Google File System, from which Hadoop is derived, all the way back in 2004. That goes to show just how long it takes for technologies used in large consumer companies like Google and Yahoo! to make their way into the enterprise.[16]

Cloudera competitor Hortonworks was spun out of Yahoo! Its engineers have contributed more than 80% of the code for Apache Hadoop. MapR focuses on delivering a higher-performance version of Hadoop with its M5 offering, which tries to address the biggest knock on Hadoop: the long time it takes to process data.

At the same time as these companies are delivering and supporting Hadoop in the enterprise, other companies are emerging to deliver Hadoop in the cloud. Qubole, Nodeable, and Platfora are three companies in the cloud Hadoop space. The challenge for these companies will be standing out from native Big Data cloud processing offerings such as Amazon's own EMR offering.

---

[14] The author is an investor in Ignition Partners.
[15] http://www.xconomy.com/san-francisco/2010/11/04/is-cloudera-the-next-oracle-ceo-mike-olson-hopes-so/2/
[16] http://static.googleusercontent.com/external_content/untrusted_dlcp/research.google.com/en/us/archive/mapreduce-osdi04.pdf

What, exactly, does Hadoop do? Hadoop was designed to perform batch-based operations across very large data sets. In the traditional Hadoop model, engineers design jobs that are processed in parallel across hundreds or thousands of servers. The individual results are pulled back together to produce the result. As a very simple example, a Hadoop MapReduce job might be used to count the number of occurrences of words in various documents. If there were millions of documents, it would be difficult to perform such a calculation on a single machine. Hadoop breaks the task into smaller jobs that each machine can execute. The results of each individual counting job are added together to produce the final count.

The challenge is that running such jobs can consume a lot of time, which is not ideal when analysts and other business users want to query data in real time. New additions to Hadoop such as the Cloudera Impala project promise to make Hadoop more responsive, not just for batch processing, but for near real-time analytics applications as well. Of course, such innovations make Cloudera a desirable acquisition target (before or after going public), either for Intel, as a follow on to its investment, or for an existing large analytics or data-warehousing provider such as EMC, HP, IBM, or Oracle, among others.[17]

■ **Note** No more waiting days or weeks for analyses to be run and interpreted. Big Data infrastructure and applications are making real-time analysis possible for the first time ever.

Big Data Applications

We use a wide variety of Big Data Applications (BDAs) every day, often without even thinking of them as BDAs. Facebook, Google, LinkedIn, Netflix, Pandora, and Twitter are just a few of the applications that use large amounts of data to give us insights and keep us entertained.

---

[17]http://blog.cloudera.com/blog/2012/10/cloudera-impala-real-time-queries-in-apache-hadoop-for-real/

While we'll continue to see innovation in Big Data infrastructure, much of the interest going forward in Big Data will be in BDAs that take advantage of the vast amounts of data being generated and the low-cost computing power available to process it. Here are a few examples:

- Facebook stores and uses Big Data in the form of user profiles, photos, messages, and advertisements. By analyzing such data, the company is better able to understand its users and figure out what content to show them.

- Google crawls billions of web pages and has a vast array of other Big Data sources such as Google Maps, which contains an immense amount of data including physical street locations, satellite imagery, on-street photos, and even inside views of many buildings.

- LinkedIn hosts hundreds of millions of online resumes as well as the knowledge about how people are connected with each other. The company uses all that data to suggest the subset of people with whom we might want to connect. LinkedIn also uses that data to show us relevant updates about our friends and colleagues.

- Internet radio service Pandora uses some 450 song attributes to figure out what songs to recommend. The company employs musicologists who characterize the attributes of virtually every new song that comes out and store its characteristics as part of the Music Genome Project.[18] The company has more than 900,000 songs from more than 90,000 artists in its database.

- Netflix is well-known for its movie prediction algorithms, which enable it to suggest to movie viewers what movie to watch next. The company relies on a group of about 40 human taggers to makes notes on more than 100 attributes, from storyline to tone, that define each movie.[19]

- Twitter handles more than 500 million tweets per day. Companies like Topsy (recently acquired by Apple), a data analytics startup that performs real-time analysis of tweets, are using such data sources to build applications on top of Twitter and other platforms.

---

[18]http://www.time.com/time/magazine/article/0,9171,1992403,00.html
[19]http://consumerist.com/2012/07/09/how-does-netflix-categorize-movies/

Big Data applications are emerging to improve our personal lives as well. Education startup Knewton uses Big Data to create dynamic learning experiences as well as to understand which students are likely to drop out so that advisors can help them stay in school. Healthcare startup Practice Fusion is improving patient experiences by making it easier for doctors to work with patient medical records. SurveyMonkey is turning responses to millions of surveys into data that markets and product designers can use to better understand their customers and clients.

Such applications are just the tip of the iceberg when it comes to Big Data Applications. Enterprises have historically built and maintained their own infrastructure for processing Big Data and in many cases developed custom applications for analyzing that data. All that is starting to change across a variety of areas, from serving existing customers better using operational intelligence tools to delivering more relevant ads online. More and more businesses will continue to take advantage of pre-built, cloud-based applications to meet their needs.

Online Advertising Applications

To determine which ad to show you, companies use algorithmic solutions to process huge volumes of data in real time. These algorithmic approaches combine a variety of rules that tell computers what to do with real-time data about what is happening in the market and data about whether you (or users similar to you) have shown an interest in related content in the past. For example, an algorithmic advertising bidding algorithm might determine that you had recently visited an automotive web site and that ad space for auto ads is available at a desirable price. Based on this information, the bidding algorithm would acquire the ad space and show you a relevant car ad.

Using this kind of automated analysis, computer programs can figure out which ad is most relevant to you and how much to pay (or charge) for particular ad impressions. Vendors in this space include Collective, DataXu, Metamarkets, Rocket Fuel, Turn, and a number of others.

These vendors truly operate at Big Data scale. The Turn platform processes some 100 billion data events daily and stores some 20 trillion attributes.[20] The Rocket Fuel platform currently handles some 43.2 billion queries per day.[21] Meanwhile, Admeld (now part of Google) works with publishers to help them optimize their ad inventory. Instead of providing just basic ad services, these companies use advanced algorithms to analyze a variety of attributes across a range of data sources to optimize ad delivery.

---

[20]http://www.turn.com/whyturn
[21]http://www.slideshare.net/HusetMarkedsforing/rtb-update-4-dominic-trigg-rocket-fuel

Marketers will continue to shift more dollars to online advertising, which suggests that this category is likely to witness growth and consolidation. Mobile advertising and mobile analytics presents one of the largest potential growth markets because of the increasingly amount of time consumers and business users now spend on their mobile devices.

Companies like Flurry (recently acquired by Yahoo!) provide analytics capabilities that allow mobile app developers to measure consumer behavior and monetize their audiences more effectively. At the same time, the mobile area is also one of the most complex due to the amount of control that Amazon, Apple, Google, and hardware vendors like Samsung exert in the space.

Sales and Marketing Applications

Salesforce.com changed the way companies did Customer Relationship Management (CRM) by introducing its hosted "no software" model for CRM as an alternative to PeopleSoft and other offerings that had to be installed, customized, and run on-premise.

More recently, marketing automation companies like Eloqua (now a part of Oracle), Marketo (now public), and HubSpot have systematized the way companies do lead management, demand generation, and email marketing.

But today's marketers face a new set of challenges. They have to manage and understand customer campaigns and interactions across a large number of channels.

Today's marketers need to ensure that a company is optimizing its web pages so they get indexed in Google and Bing and are easy for potential customers to find. They need to have a regular presence on social media channels such as Facebook, Twitter, and Google Plus. This is not just because these are the venues where people are spending time getting entertained and receiving information, but also because of Google's increasing emphasis on social media as a way to gauge the importance of a particular piece of content.

As Patrick Moran, Chief Marketing Officer at application performance monitoring company New Relic, points out, marketers also need to factor in a variety of other sources of data to understand their customers fully. This includes actual product usage data, lead sources, and trouble ticket information. Such data can give marketers significant insight into which customers are most valuable so they can look for other potential customers with similar attributes. They can also determine which marketing activities are most likely to result in prospects converting into customers.

All of this means a lot of data for marketers to visualize and act on. As Brian Kardon, chief marketing officer of Lattice Engines, and formerly of Eloqua and Forrester Research, suggests, marketing in the future will be in large part about algorithms. Trading on Wall Street was once the purview of humans, until computer-run algorithmic trading took its place. Kardon envisions a similar future for marketing, a future in which algorithms analyze all of these data sources to find useful patterns and tell marketers what to do next.[22]

■ **Note** Cloud-based sales and marketing software now allows companies to analyze any bit of data streaming into the company, including product usage, online behavior, trouble ticket information, and more. Marketers need to be on their toes to implement the technologies necessary to use such data before competitors do.

Such software will likely tell marketers which campaigns to run, which emails to send, what blog posts to write, and when and what to tweet. It won't stop there, however.

Ultimately, Big Data marketing applications will not only analyze all these data sources but perform much of the work to optimize marketing campaigns based on the data. Companies like BloomReach and Content Analytics are already heading down this path with algorithm-based software that helps e-commerce companies optimize their web sites for highest conversion and visibility.

Of course, the creative part of marketing will remain critical, and marketers will still make the big picture decisions about where to invest and how to position relative to their competition. But BDAs will play a huge role in automating much of the manual work currently associated with online marketing.

Visualization Applications

As access to data becomes more democratized, visualization becomes ever more important. There are many companies in the visualization space, so in this section we'll highlight just a few of them. Tableau Software is well known for its interactive and easy-to-use visualization software. The company's technology came out of research at Stanford University. The company recently completed a successful public offering.

QlikTech offers its popular QlikView visualization product, which some 26,000 companies use around the world. The company went public in 2010 and is valued at about $2.3 billion as of February, 2014. There are a host of other recent entrants. SiSense performs huge amounts of data crunching on computers as small as laptops and delivers visualizations as output.

---

[22]http://www.b2bmarketinginsider.com/strategy/real-time-marketing-trading-room-floor

While not strictly a visualization company, Palantir is well known for its Big Data software and has a strong customer base in government and financial services. There are also offerings from large enterprise vendors, including IBM, Microsoft, Oracle, SAS, SAP, and TIBCO.

More and more companies are adding tools for embedding interactive visualizations into web sites. Publishers now use such visualizations to provide readers with greater insights into data.

Enterprise collaboration and social networking companies like Yammer emerged to make business communication, both internally and externally, more social. Expect to see similar social capabilities become a standard part of nearly every data analytics and visualization offering.

Given the importance of visualization as a way to understand large data sets and complex relationships, new tools will continue to emerge. The challenge and opportunity for such tools is to help people see data in ways that provide greater insights, as well as to help them take meaningful action based on those insights.

Business Intelligence Applications

Much of the history of data analysis has been in business intelligence (BI). Organizations rely on BI to organize and analyze large quantities of corporate data with the goal of helping managers make better decisions. For example, by analyzing sales and supply chain data, managers might be able to decide on better pricing approaches in the future.

Business intelligence was first referenced in a 1958 article by an IBM researcher, and the company has continued to break new ground with technical advances like algorithmic trading and IBM Watson. Other major vendors, including SAP, SAS, and Oracle, all offer business intelligence products. MicroStrategy remains one of the few large-scale, independent players in the space. It has a market cap of about a billion dollars and has recently introduced compelling mobile offerings to the market.[23]

Domo, a cloud-based business intelligence software company, is a relatively recent entrant into the market. Domo was founded by Josh James, the founder and former CEO of analytics leader Omniture (now Adobe). Other well-known players in the space include GoodData and Birst. However, the hosted BI space has proven difficult to crack due to the challenge of getting companies to move mission-critical company data into the cloud and the amount of

---

[23]http://en.wikipedia.org/wiki/Business_intelligence

customization traditionally required. But as with other areas, that is starting to change. Users want to be able to access and work with the same data on their desktops, tablets, and mobile phones, which means that more data will need to move into the cloud for it to be easily accessed and shared.

Operational Intelligence

By performing searches and looking at charts, companies can understand the cause of server failures and other infrastructure issues. Rather than building their own scripts and software to understand infrastructure failures, enterprises are starting to rely on newer operational intelligence companies like Splunk. The company provides both an on-premise and cloud-based version of its software, which IT engineers use to analyze the vast amounts of log data that servers, networking equipment, and other devices generate.

Sumo Logic and Loggly are more recent entrants in the space, and large vendors such as TIBCO (which acquired LogLogic) and HP (which acquired ArcSight) and existing player Vitria all have offerings as well. Meanwhile, AppDynamics and New Relic provide web offerings that enable engineers to understand issues at the application layer.

Data as a Service

Straddling both Big Data infrastructure and applications is the Data as a Service category. Historically, companies have faced a challenge in obtaining Big Data sets. It was often hard to get up-to-date data or to get it over the Internet. Now, however, data as a service providers come in a variety of forms. Dun & Bradstreet provides web programming interfaces for financial, address, and other forms of data, while FICO provides financial data. Huge volumes of streaming data, such as the tweet streams that Twitter provides, are among the most interesting new data sources.

These data sources allow others to build compelling applications based on them. New applications can predict the outcomes of presidential elections with very high accuracy or help companies understand how large numbers of consumers feel about their brands. There are also companies that deliver vertical-specific data through service interfaces. BlueKai provides data related to consumer profiles, INRIX provides traffic data, and LexisNexis provides legal data.

Data Cleansing

Perhaps one of the most unglamorous yet critical areas when it comes to working with data is that of data cleansing and integration. Companies like Informatica have long played a dominant role in this space. Internal and external data can be stored in a wide range of formats and can include errors and duplicate records. Such data often needs to be cleansed before it can be used or before multiple data sources can be used together.

At its simplest level, data cleansing involves tasks like removing duplicate records and normalizing address fields. Companies like Trifacta, funded by well-known venture capital firms Andreessen Horowitz and Greylock, are bringing to market new applications for cleansing data and working with diverse data sources. Data cleansing as a cloud service, perhaps through a combination of machine-based algorithms and crowdsourcing, presents an interesting and as yet mostly unexplored market opportunity.

Data Privacy

As we move more data to the cloud and publish more information about ourselves on the net, data privacy and security remains a growing concern. Facebook has beefed up the control that users have over which information they share. Google has added layers of security and encryption to make its data centers more secure. Recent large-scale losses of consumer data at retailers like Target have also raised awareness around data security and privacy for both consumers and businesses.

As with any new technology, Big Data offers many potential benefits, but it also poses a variety of risks. It brings with it the need for well-thought-out policies around the kinds of data we store and how we use that data. As an example, anonymous data may not be as anonymous as it seems. In one study, analysts were able to look at anonymized movie-watching data and determine, by looking at reviews from users who had posted on the Internet Movie Database (IMDB), which users had watched which movies.

In the future, BDAs may emerge that not only let us decide which data to share, but also help us to understand the hidden implications of sharing personal information, whether that information identifies us personally or not.

Landscape Futures

Data and the algorithms designed to make use of it are becoming a fundamental and distinguishing competitive asset for companies, both consumer- and business-focused.

File sharing and collaboration solutions Box and Dropbox may well be considered BDAs given the huge volume of files they store. More and more BDAs are emerging that are vertical focused. As mentioned in chapters to come, those like Opower take the data from power meters and help consumers and businesses understand their power consumption and then use energy more efficiently.

Nest, which was recently acquired by Google for $3.2 billion, is a learning thermostat that understands consumer behavior and applies algorithms to the data it collects so that it can better heat and cool homes. Some of these BDA companies will be acquired, while others will follow in the footsteps of QlikTech, Tableau, and Splunk and go public as they seek to build leading, independent BDA companies. What's clear is that the market's appetite for Big Data is just beginning.

■ **Note**   The market's appetite for Big Data is just beginning.

As more BDAs come to market, where does that leave infrastructure providers? When it comes to cloud-based infrastructure, it's likely that Amazon will have a highly competitive offering in virtually every area. And for those areas where it doesn't, the larger open source infrastructure vendors may jump in to provide cloud-based offerings. If history is any predictor, it's likely that big enterprise players, from Cisco to EMC, IBM, and Oracle, will continue to be extremely active acquirers of these vendors. Intel's $900 million investment in Cloudera shows that it is extremely serious about Big Data.

Going forward, expect even more BDAs to emerge that enhance our work and our personal lives. Some of those applications will help us better understand information. But many of them won't stop there. BDAs will improve our ability to reach the right customers and to serve those customers effectively once we do. But they won't stop there. BDAs at the intersection of Big Data and mobile will improve the quality of care we receive when we visit the doctor's office. Someday, BDAs may even enable self-driving cars to take care of our daily commutes. No matter the application, the Big Data landscape is sure to be a source of profound innovation in the years ahead.

Your Big Data Roadmap

Big Data: Where to Start

Thinking about how you want to act on your results after your data gathering and analysis is complete will help you navigate your way to a successful Big Data outcome.

In some cases, the results of your project may be interactive visualizations or dashboards you present to management, partners, or customers. In other cases, you might implement automated systems that use Big Data to take algorithmic action—to make better financial decisions, change pricing automatically, or deliver more targeted ads.

Your roadmap should also include a plan for pulling together the right team members and for getting access to the necessary data assets you want to analyze. With your vision and key questions in hand, you can then combine the necessary people, technology, and data sources to deliver the answers you need.

Goodbye SQL, Hello NoSQL

First, let's take a look at one of the highest profile, emerging areas in Big Data—that of SQL and NoSQL. In many ways, NoSQL data stores are a step back to the future.

Traditional Relational Database Management Systems (RDBMS) systems rely on a table-based data store and a structured query language (SQL) for accessing data. In contrast, NoSQL systems do not use a table-based approach, nor do they use SQL. Instead, they rely on a key-value store approach to data storage and lookup. NoSQL systems are actually a lot like the Information Management Systems (IMS) that were commonly used before the introduction of relational database systems!

Before relational database systems, data was stored in relatively simple data stores, not unlike the key-value systems of today's NoSQL offerings. What has changed is that, after a very long period (some 40 years) of using computer systems to store structured data, unstructured data is now growing at a much faster rate than its structured counterpart. The IMS is back in favor and it is known as NoSQL. What has also changed is that just as MySQL became popular due to its open source distribution approach, NoSQL is benefiting from open source popularity as well, primarily in the form of MongoDB and Cassandra.

Like the MySQL database, which was open source but had commercial support available through the company of the same name, MongoDB takes a similar approach. MongoDB is available as free, open source software, with commercial versions and support available from the company of the same name. Cassandra is another NoSQL database, and it's maintained by the Apache Software Foundation.

MongoDB is the most popular NoSQL database and MongoDB (the company) has raised some $231 million in funding. DataStax offers commercial support for Cassandra and has raised some $83 million to date. Aerospike is another player in the field and has focused on improving NoSQL performance by optimizing its offering specifically to take advantage of the characteristics of flash-based storage. Aerospike recently decided to open-source its database server code.

NoSQL databases are particularly well-suited as document stores. Traditional structured databases require the up-front definition of tables and columns before they can store data. In contrast, NoSQL databases use a dynamic schema that can be changed on the fly.

Due to the huge volume of unstructured data now being created, NoSQL databases are surging in popularity. It may be a little too soon to say "goodbye" to SQL databases, but for applications that need to store vast quantities of unstructured data, NoSQL makes a lot of sense.

Compiling and Querying Big Data

Identifying the data sources you want to work with is one of the first major steps in developing your Big Data roadmap. Some data may be public, such as market, population, or weather data. Other sources of data may be in a variety of different locations internally. If you're looking to analyze and reduce network security issues, your data may be in the form of log files spread out across different network devices.

If your goal is to increase revenue through better sales and marketing, your data may be in the form of web site logs, application dashboards, and various analytics products, both in-house and cloud-based. Or you may be working on a Big Data legal compliance project. In this case, your data is in the form of documents and emails spread across email systems, file servers, and system backups. Financial transaction data may be stored in different data warehouses.

Depending on the tools you choose to use, you may not need to pull all the data together into one repository. Some products can work with the data where it resides, rather than pulling it into a common store. In other cases, you may want to compile all the data into one place so that you can analyze it more quickly.

Once you have access, your next step is to run queries to get the specific data you need. Despite the growing size of unstructured data, the vast majority of Big Data is still queried using SQL and SQL-like approaches. But first, let's talk about how to get the data into a state you can work with.

One of the biggest issues that typically comes up when working with Big Data is data formatting For example, log files produced by network and computer systems manufactured by different vendors contain similar information but in different formats. For example, all the log files might contain information about what errors occurred, when those errors occurred, and on which device.

But each vendor outputs such error information in a slightly different format. The text identifying the name of a device might be called DeviceName in one log file and device_name in another. Times may be stored in different formats; some systems store time in Universal Coordinated Time (UTC), whereas other systems record time in local time, say Pacific Standard Time (PST) if that is the time zone in which a particular server or device is running.

Companies like Informatica have historically developed tools and services to address this issue. However, such approaches require users to define complex rules to transform data into consistent formats. More recently, companies like Trifacta, founded by UC Berkeley Computer Science professor Joe Hellerstein, have sought to simplify the data-transformation challenge through modeling and machine learning software.

Once you have the data in a workable state, your next step is to query the data so you can get the information you need. For structured data stored in Oracle, DB2, MySQL, PostgreSQL, and other structured databases, you'll use SQL. Simple commands like SELECT allow you to retrieve data from tables. When you want to combine data from multiple tables you can use the JOIN command. SQL queries can get quite complex when working with many tables and columns, and expert database administrators can often optimize queries to run more efficiently. In data-intensive applications, quite often the data store and the queries used to access the data stored therein can become the most critical part of the system.

While SQL queries can sometimes be slow to execute, they are typically faster than most Hadoop-based Big Data jobs, which are batch-based. Hadoop, along with MapReduce, has become synonymous with Big Data for its ability to distribute massive Big Data workloads across many commodity servers. But the problem for data analysts used to working with SQL was that Hadoop did not support SQL-style queries, and it took a long time to get results due to Hadoop's batch-oriented nature.

■ **Note** While traditional SQL-oriented databases have until recently provided the easiest way to query data, NoSQL technology is catching up fast. Apache Hive, for example, provides a SQL-like interface to Hadoop.

Several technologies have evolved to address the problem. First, Hive is essentially SQL for Hadoop. If you're working with Hadoop and need a query interface for accessing the data, Hive, which was originally developed at Facebook, can provide that interface. Although it isn't the same as SQL, the query approach is similar.

After developing your queries, you'll ultimately want to display your data in visual form. Dashboards and visualizations provide the answer. Some data analysis and presentation can be done in classic spreadsheet applications like Excel. But for interactive and more advanced visualizations, it is worth checking out products from vendors like Tableau and Qliktech. Their products work with query languages, databases, and both structured and unstructured data sources to convert raw analytics results into meaningful insights.

Big Data Analysis: Getting What You Want

The key to getting what you want with Big Data is to adopt an iterative, test-driven approach. Rather than assuming that a particular data analysis is correct, you can draw an informed conclusion and then test to see if it's correct.

For example, suppose you want to apply Big Data to your marketing efforts, specifically to understand how to optimize your lead-generation channels. If you're using a range of different channels, like social, online advertising, blogging, and search engine optimization, you'll get the data about all those sources into one place. After doing that, you'll analyze the results and determine the conversion rates from campaign to lead to prospect to customer, by channel. You may even take things a step further, producing a granular analysis that tells you what time of day, day of week, kinds of tweets, or types of content produce the best results.

From this analysis, you might decide to invest more in social campaigns on a particular day of the week, say Tuesdays. The great news with Big Data is that you can iteratively test your conclusions to see if they are correct. In this case, by running more social media marketing campaigns on Tuesdays, you'll quickly know whether that additional investment makes sense. You can even run multivariate tests—combining your additional social media investment with a variety of different kinds of content marketing to see which combinations convert the most prospects into customers.

Because Big Data makes it more cost-effective to collect data and faster to analyze that data, you can afford to be a lot more iterative and experimental than in the past. That holds true whether you're applying Big Data to marketing, sales, or to virtually any other part of your business.

Big Data Analytics: Interpreting What You Get

When it comes to interpreting the results you get from Big Data, context is everything. It's all too easy to take data from a single point in time and assume that those are representative. What really matters with Big Data is looking at the trend.

Continuing with the earlier example, if lead conversions from one channel are lower on one day of the week but higher on another, it would be all too easy to assume that those conversion rates will remain the same over time. Similarly, if there are a few errors reported by a computer or piece of network equipment, those might be isolated incidents. But if those errors keep happening or happen at the same time as other errors in a larger network, that could be indicative of a much larger problem.

In reality, it's crucial to have data points over time and in the context of other events. To determine if a particular day of the week produces better results from social marketing efforts than other days of the week, you need to measure the performance of each of those days over multiple weeks and compare the results. To determine if a system error is indicative of a larger network failure or security attack, you need to be able to look at errors across the entire system.

Perhaps no tool is more powerful for helping to interpret Big Data results than visualization. Geographic visualizations can highlight errors that are occurring in different locations. Time-based visualizations can show trends, such as conversion rates, revenue, leads, and other metrics over time. Putting geographic and time-based data together can produce some of the most powerful visualizations of all. Interactive visualizations can enable you to go backward and forward in time. Such approaches can be applied not only to business but also to education, healthcare, and the public sector. For example, interactive visualizations can show changes in population and GDP on a country-by-country basis over time, which can then be used to evaluate potential investments in those regions going forward.

You'll learn about visualization in detail later in the book, but at a high level, you can use tools from vendors like Tableau and Qliktech's Qlikview to create general visualizations. For specific areas, such as geographic visualizations, products like CartoDB make it easy to build visualizations into any web page with just a few lines of HTML and JScript.

In some cases, interpreting the data means little to no human interaction at all. For online advertising, financial trading, and pricing, software combined with algorithms can help you interpret the results and try new experiments. Computer programs interpret complex data sets and serve up new advertise-ments, trade stocks, and increase or decrease pricing based on what's working and what's not. Such automated interpretation is likely to take on a larger and larger role as data sets continue to grow in volume and decisions have to be made faster and on a larger scale than ever before. In these systems, human oversight is critical—dashboards that show changes in system behavior, highlight exceptions, and allow for manual intervention are key.

Should I Throw Out My RDBMS?

Practically since they were first invented, people have been trying to get rid of databases. The simple answer is that you shouldn't throw out your RDBMS. That's because each Big Data technology has its place, depending on the kind of data you need to store and the kind of access you need to that data.

Relational database systems were invented by Edgar Codd at IBM in the late 1960s and early 1970s. Unlike its predecessor, the Information Management System (IMS), an RDBMS consists of various tables. In contrast, IMS systems are organized in a hierarchical structure and are often used for recording high-velocity transaction data, such as financial services transactions. Just as radio did not disappear when television came into existence, IMS systems did not disappear when RDBMS appeared. Similarly, with newer technologies like Hadoop and MapReduce, the RDBMS is not going away.

First, it would be impractical simply to get rid of classic structured database systems like Oracle, IBM DB2, MySQL, and others. Software engineers know these systems well and have worked with them for years. They have written numerous applications that interface with these systems using the well-known structured query language or SQL. And such systems have evolved over time to support data replication, redundancy, backup, and other enterprise requirements.

Second, the RDBMS continues to serve a very practical purpose. As implied by SQL, the language used to store data to and retrieve data from databases, an RDBMS is most suitable for storing structured data. Account information, customer records, order information, and the like are typically stored in an RDBMS.

But when it comes to storing unstructured information such as documents, web pages, text, images, and videos, using an RDBMS is likely not the best approach. RDBMS systems are also not ideal if the structure of the data you need to store changes frequently.

In contrast to traditional database systems, the more recently introduced Big Data technologies—while excelling at storing massive amounts of unstructured data—do not do as well with highly structured data. They lack the elegant simplicity of a structured query language. So while storing data is easy, getting it back in the form you need is hard.

That's why today's Big Data systems are starting to take on more of the capabilities of traditional database systems, while traditional database systems are taking on the characteristics of today's Big Data technologies. Two technology approaches at opposite ends of the spectrum are becoming more alike because customers demand it.

Today's users aren't willing to give up the capabilities they've grown accustomed to over the past 40 years. Instead, they want the best of both worlds—systems that can handle both structured and unstructured data, with the ability to spread that data over many different machines for storage and processing.

■ **Note** Big Data systems are less and less either SQL or NoSQL. Instead, customers more and more want the best of both worlds, and vendors as a result are providing hybrid products sporting the strengths of each.

Big Data Hardware

One of the big promises of Big Data is that it can process immense amounts of data on commodity hardware. That means companies can store and process more data at a lower cost than ever before. But it's important to keep in mind that the most data-intensive companies companies like Facebook, Google, and Amazon are designing their own custom servers to run today's most data-intensive applications.

So when it comes to designing your Big Data roadmap, there are a few different approaches you can take to your underlying infrastructure—the hardware that runs all the software that turns all that data into actionable insights.

One option is to build your own infrastructure using commodity hardware. This typically works well for in-house Hadoop clusters, where companies want to maintain their own infrastructure, from the bare metal up to the application. Such an approach means it's easy to add more computing capacity to your cluster—you just add more servers. But the hidden cost plays out over time. While such clusters are relatively easy to set up, you're on the book to manage both the underlying hardware and the application software.

This approach also requires you to have hardware, software, and analytics expertise in-house. If you have the resources, this approach can make the most sense when experimenting with Big Data as a prototype for a larger application. It gives you first-hand knowledge of what is required to set up and operate a Big Data system and gives you the most control over underlying hardware that often has a huge impact on application performance.

Another option is to go with a managed service provider. In this case, your hardware can reside either in-house or at the data center of your provider. In most cases, the provider is responsible for managing hardware failures, power, cooling, and other hardware-related infrastructure needs.

A still further option is to do all your Big Data processing in the cloud. This approach allows you to spin up and down servers as you need them, without making the up-front investment typically required to build your own Big Data infrastructure from scratch. The downside is there's often a layer of virtualization software between your application and the underlying hardware, which can slow down application performance.

■ **Note** Cloud technology allows you to set up servers as necessary and scale the system quickly—and all without investing in your own hardware. However, the virtualization software used in many cloud-based systems slows things down a bit.

In days past IBM, Sun Microsystems, Dell, HP, and others competed to offer a variety of different hardware solutions. That is no longer the case. These days, it's a lot more about the software that runs on the hardware rather than the hardware itself.

A few things have changed. First, designing and manufacturing custom hardware is incredibly expensive and the margins are thin. Second, the scale-out (instead of scale-up) nature of many of today's applications means that when a customer needs more power or storage, it's as simple as adding another server. Plus, due to Moore's Law, more computing power can fit in an ever-smaller space. By packing multiple cores (processors) into each processor, each processor packs more punch. Each server has more memory, and each drive stores more data.

There's no doubt about it, though. Today's Big Data applications are memory and storage hogs. In most cases, you should load up your servers, whether they are local or cloud-based, with as much memory as you can. Relative to memory, disks are still relatively slow. As a result, every time a computer has to go to disk to fetch a piece of data, performance suffers. The more data that can be cached in memory, the faster your application—and therefore your data processing—will be.

The other key is to write distributed applications that take advantage of the scale-out hardware approach. This is the approach that companies like Facebook, Google, Amazon, and Twitter have used to build their applications. It's the approach you should take so that you can easily add more computing capacity to meet the demands of your applications. Every day more data is created, and once you get going with your own Big Data solutions, it's nearly a sure thing that you'll want more storage space and faster data processing.

Once your Big Data project becomes large enough, it may make sense to go with hardware specifically optimized for your application. Vendors like Oracle and IBM serve some of the largest applications. Most of their systems are built from commodity hardware parts like Intel processors, standard memory, and so on. But for applications where you need guaranteed reliability and service you can count on, you may decide it's worth paying the price for their integrated hardware, software, and service offerings.

Until then, however, commodity, off-the-shelf servers should give you everything you need to get started with Big Data. The beauty of scale-out applications when it comes to Big Data is that if a server or hard drive fails, you can simply replace it with another. But definitely buy more memory and bigger disks than you think you'll need. Your data will grow a lot faster and chew up a lot more memory than your best estimates can anticipate.

Data Scientists and Data Admins: Big Data Managers

Consulting firm McKinsey estimates that Big Data can have a huge impact in many areas, from healthcare to manufacturing to the public sector. In healthcare alone, the United States could reduce healthcare expenditure some 8% annually with effective use of Big Data.

Yet we are facing a shortage of those well-versed in Big Data technologies, analytics, and visualization tools. According to McKinsey, the United States alone faces a shortage of 140,000 to 190,000 data analysts and 1.5 million "managers and analysts with the skills to understand and make decisions based on the analysis of Big Data."[1]

There is no standard definition of the roles of data scientist and data analyst. However, data scientists tend to be more versed in the tools required to work with data, while data analysts focus more on the desired outcome—the questions that need to be answered, the business objectives, the key metrics, and the resulting dashboards that will be used on a daily basis. Because of the relative newness of Big Data, such roles often blur.

When it comes to working with structured data stored in a traditional database, there is nothing better than having an experienced database administrator (DBA) on hand. Although building basic SQL queries is straightforward, when working with numerous tables and lots of rows, and when working with data stored in multiple databases, queries can grow complex very quickly. A DBA can figure out how to optimize queries so that they run as efficiently as possible.

Meanwhile, system administrators are typically responsible for managing data repositories and sometimes for managing the underlying systems required to store and process large quantities of data. As organizations continue to capture and store more data—from web sites, applications, customers, and physical systems such as airplane engines, cars, and all of the other devices making up the Internet of Things—the role of system and database administrators will continue to grow in importance. Architecting the right systems from the outset and choosing where and how to store the data locally or in the cloud is critical to a successful long-term Big Data strategy. Such a strategy is the basis for building a proprietary Big Data asset that delivers a strategic competitive advantage.

All of these roles are critical to getting a successful outcome from Big Data. All the world's data won't give you the answers you need if you don't know

---

[1]McKinsey & Company Interactive, Big Data: The Next Frontier for Competition. www.mckinsey.com/features/big_data

what questions to ask. Once the answers are available, they need to be put into a form that people can act on. That means translating complex analysis, decision trees, and raw statistics into easy-to-understand dashboards and visualizations.

The good news for those without the depth of skills required to become Big Data experts is that dozens of Big Data related courses are now available online, both from companies and at local colleges and universities. Vendors like IBM and Rackspace also offer Big Data courses online, with Rackspace recently introducing its free-to-view CloudU.[2]

One Fortune 500 company I work with runs an internal university as part of its on-going human capital investment. They have recently added Big Data, cloud, and cyber-security courses from my firm—The Big Data Group—to their offerings. In these courses, we not only cover the latest developments in these areas but also discuss Big Data opportunities the participants can explore when they return to their day to day jobs.

Given the numbers cited by the McKinsey report, I expect that more companies will invest in Big Data, analytics, and data administration education for executives and employees alike in the months and years ahead. That's good news for those looking to break into Big Data, change job functions, or simply make better use of data when it comes to making important business decisions.

Chief Data Officer: Big Data Owner

According to industry research firm Gartner, by 2015, 25% of large global organizations will have appointed a chief data officer (CDO).[3] The number of CDOs in large organizations doubled from 2012 to 2014. Why the rise of the CDO?

For many years, information technology (IT) was primarily about hardware, software, and managing the resources, vendors, and budgets required to provide technology capabilities to large organizations. A company's IT resources were a source of competitive advantage. The right IT investments could enable an organization to be more collaborative and nimble than its competitors. Data was simply one component of a larger IT strategy, quite often not even thought about in its own right separate from storage and compute resources.

---

[2]http://www.rackspace.com/blog/bigger-is-better-introducing-the-cloudu-big-data-mooc/
[3]http://www.gartner.com/newsroom/id/2659215

Now data itself is becoming the source of strategic competitive advantage that IT once was, with storage hardware and computing resources the supporting actors in the Big Data play.

At the same time that companies are looking to gain more advantage from Big Data, the uses and misuses of that data has come under higher scrutiny. The dramatic rise in the amount of information that companies are storing, combined with a rising number of cases in which hackers have obtained personal data companies are storing has put data into its own class.

2013 saw two very high-profile data breaches. Hackers stole more than 70 million personal data records from retailer Target that included the name, address, email address, and phone number information of Target shoppers. Hackers also stole some 38 million user account records from software maker Adobe. Then in 2014, hackers gained access to eBay accounts, putting some 145 million user accounts at risk.

The prominence of data has also increased due to privacy issues. Customers are sharing a lot more personal information than they used to, from credit card numbers to buying preferences, from personal photos to social connections. As a result, companies know a lot more about their customers than they did in years past. Today, just visiting a web site creates a huge trail of digital information. Which products did you spend more time looking at? Which ones did you skip over? What products have you looked at on other web sites that might be related to your purchase decision? All that data is available in the form of digital breadcrumbs that chart your path as you browse the web, whether on your computer, tablet, or mobile phone.

Demand for access to data—not just to the data itself but to the right tools for analyzing it, visualizing it, and sharing it—has also risen. Once the domain of data analysis experts versed in statistics tools like R, actionable insights based on data are now in high demand from all members of the organization. Operations managers, line managers, marketers, sales people, and customer service people all want better access to data so they can better serve customers.

What's more, the relevant data is just as likely to be in the cloud as it is to be stored on a company's in-house servers. Who can pull all that data together, make sure the uses of that data adhere to privacy requirements, and provide the necessary tools, training, and access to data to enable people to make sense of it? The CDO.

IT was and continues to be one form of competitive advantage. Data is now rising in importance as another form of competitive advantage—a strategic asset to be built and managed in its own right. As a result, I expect many more organizations to elevate the importance of data to the executive level in the years ahead.

Your Big Data Vision

To realize your Big Data vision, you'll need the right team members, technology, and data sources. Before you assemble your team, you might want to experiment with a few free, online Big Data resources. The Google Public Data Explorer lets you interact with a variety of different public data sources at `www.google.com/publicdata/directory`. Even though these may not be the data sources you eventually want to work with, spending a few hours or even a few minutes playing around with the Data Explorer can give you a great sense for the kinds of visualizations and dashboards you might want to build to help you interpret and explain your Big Data results.

With your Big Data outcome in mind, you can begin to assemble your team, technology, and data resources.

Your team consists of both actual and virtual team members. Bring in business stakeholders from different functional areas as you put your Big Data effort together. Even though your project may be focused just on marketing or web site design, it's likely you'll need resources from IT and approval from legal or compliance before your project is done.

So many different Big Data technologies are available that the choice can often be overwhelming. Don't let the technology drive your project—your business requirements should drive your choice of technology.

Finally, you may not be able to get access to all the data you need right away. This is where having the dashboards you ultimately want to deliver designed and sketched out up front can help keep things on track. Even when you can't fill in every chart, using mock-up visualizations can help make it clear to everyone which data sources are most important in delivering on your Big Data vision.

In the chapters ahead, we'll explore a variety of different technologies, visualization tools, and data types in detail. Starting with the end in mind and taking an iterative, test-driven approach will help turn your Big Data vision into reality.

Big Data at Work

How Data Informs Design

When it comes to data and design, there's a struggle going on. Some believe that the best designs come from gut instinct while others hold that design should be data-driven. Data does have a role to play in design and modern technologies can provide deep insight into how users interact with products—which features they use and which ones they don't. This chapter explores how some of today's leading companies, including Facebook and Apple, apply data to design and how you can bring a data-informed design approach to developing your own products.

How Data Informs Design at Facebook

If there is one company whose design decisions impact a lot of people, more than a billion of them, it's social networking giant Facebook. Quite often, when Facebook makes design changes, users don't like them. In fact, they hate them.

When Facebook first rolled out its News Feed feature in 2006, back when the social networking site had just 8 million users, hundreds of thousands of students protested. Yet the News Feed went on to become one of the site's[1] most popular features, and the primary driver of traffic and engagement, according to Facebook Director of Product Adam Mosseri.[2]

That's one of the reasons that Facebook takes what Mosseri refers to as a data-informed approach, not a data-driven approach to decision making. As Mosseri points out, there are a lot of competing factors that can inform

---

[1]http://www.time.com/time/nation/article/0,8599,1532225,00.html
[2]http://uxweek.com/2010/videos/video-2010-adam-mosseri

product design decisions. Mosseri highlights six such factors: quantitative data, qualitative data, strategic interests, user interests, network interests, and business interests.

■ **Tip** Rather than taking a strict *data-driven approach* to product decisions, consider taking a more flexible *data-informed approach*. That distinction has served Facebook well.

Quantitative data is the kind of data that shows how people actually use the Facebook product. This might be, for example, the percentage of users who upload photos or the percentage who upload multiple photos at a time instead of just one.

According to Mosseri, 20% of Facebook's users—those who log in more than 25 days per month—generate 85% of the site's content. So getting just a few more people to generate content on the site, such as uploading photos, is incredibly important.

Qualitative data is data like the results from eye-tracking studies. An eye-tracking study watches where your eyes go when you look at a web page. Eye-tracking studies give a product designer critical information about whether elements of a web page are discoverable and whether information is presented in a useful manner. Studies can present viewers with two or more different designs and see which design results in more information retention, which is important knowledge for designing digital books or building a compelling news site, for example.[3]

Mosseri highlights Facebook's introduction of its Questions offering, the capability to pose a question to friends and get an answer, as an example of a *strategic interest*. Such interests might compete with or highly impact other interests. In the case of Questions, the input field necessary to ask a question would have a strong impact on the status field that asks users, "What's on your mind?"

User interests are the things people complain about and the features and capabilities they ask for.

Network interests consist of factors such as competition as well as regulatory issues that privacy groups or the government raise. Facebook had to incorporate input from the European Union for its Places features, for example.

---

[3]http://www.ojr.org/ojr/stories/070312ruel/

Finally, there are *business interests*, which are elements that impact revenue generation and profitability. Revenue generation might compete with user growth and engagement. More ads on the site might produce more revenue in the short term but at the price of reduced engagement in the long term.

One of the challenges with making exclusively data-driven decisions, Mosseri points out, is the risk of optimizing for a local maximum. He cites two such cases at Facebook: photos and applications.

Facebook's original photo uploader was a downloadable piece of software that users had to install in their web browsers. On the Macintosh Safari browser, users got a scary warning that said, "An applet from Facebook is requesting access to your computer." In the Internet Explorer browser, users had to download an ActiveX control, a piece of software that runs inside the browser. But to install the control, they first had to find and interact with an 11-pixel-high yellow bar alerting them about the existence of the control.

The design team found that of the 1.2 million people that Facebook asked to install the uploader, only 37 percent tried to do so. Some users already had the uploader installed, but many did not. So as much as Facebook tried to optimize the photo uploader experience, the design team really had to revisit the entire photo-uploading process. They had to make the entire process a lot easier—not incrementally better, but significantly better. In this case, the data could indicate to Facebook that they had an issue with photo uploads and help with incremental improvement, but it couldn't lead the team to a new design based on a completely new uploader.

The visual framework that Facebook implemented to support third-party games and applications, known as Facebook applications, is another area the design team had to redesign completely. With well-known games like Mafia Wars and FrontierVille (both created by Zynga), the navigation framework that Facebook implemented on the site inherently limited the amount of traffic it could send to such applications. While the design team was able to make incremental improvements within the context of the existing layout, they couldn't make a significant impact. It took a new layout to produce a dramatic uplift in the amount of traffic such applications received.

As Mosseri puts it, "real innovation invariably involves disruption." Such disruptions, like the News Feed, often involve a short-term dip in metrics, but they are the kinds of activities that produce meaningful long-term results. When it comes to design at Facebook, data informs design; it doesn't dictate it.

Mosseri highlights one other point about how Facebook has historically done design: "We've gotten away with a lot of designing for ourselves." If that sounds familiar, it's because it's the way that another famous technology company designs its products too.

Apple Defines Design

If there is one company that epitomizes great design, it's Apple. As Steve Jobs once famously said, "We do not do market research."[4] Rather, said Jobs, "We figure out what we want. And I think we're pretty good at having the right discipline to think through whether a lot of other people are going to want it too."

When it comes to the world's most famous design company, a few things stand out, according to Michael Lopp, former senior engineering manager at Apple, and John Gruber.[5]

First, Apple thinks good design is a present. Apple doesn't just focus on the design of the product itself, but on the design of the package the product comes in. "The build up of anticipation leading to the opening of the present that Apple offers is an important—if not the most important—aspect of the enjoyment people derive from Apple's products." For Apple, each product is a gift within a gift within a gift: from the package itself to the look and feel of the iPad, iPhone, or MacBook, to the software that runs inside.

Next, "pixel-perfect mockups are critical." Apple designers mock up potential designs down to the very last pixel. This approach removes all ambiguity from what the product will actually look like. Apple designers even use real text rather than the usual Latin "lorem ipsum" text found in so many mockups.

Third, Apple designers typically make 10 designs for any potential new feature. Apple design teams then narrow the 10 designs down to three and then down to one. This is known as the 10:3:1 design approach.

Fourth, Apple design teams have two kinds of meetings each week. Brainstorm meetings allow for free thinking with no constraints on what can be done or built. Production meetings focus on the reality of structure and the schedules required to ship products.

Apple does a few other things that set its design approach apart from others as well. The company doesn't do market research. Instead employees focus on building products that they themselves would like to use.[6]

---

[4] http://money.cnn.com/galleries/2008/fortune/0803/gallery.jobsqna.fortune/3.html
[5] http://www.pragmaticmarketing.com//resources/you-cant-innovate-like-apple?p=1
[6] This myth has been dispelled, at least to some extent, by documents made public as a result of the Apple-Samsung court case, citing a recent market research study the company conducted; see http://blogs.wsj.com/digits/2012/07/26/turns-out-apple-conducts-market-research-after-all/.

The company relies on a very small team to design its products. Jonathan Ive, Apple's Senior Vice President of Industrial Design, relies on a team of no more than 20 people to design most of Apple's core products.

Apple owns both the hardware and software, making it possible to deliver a fully integrated, best-of-breed experience. What's more, the company focuses on delivering a very small number of products for a company its size. This allows the company to focus on delivering only best-of-breed products. Finally, the company has "a maniacal focus on perfection," and Jobs was said to dedicate half his week to the "high- and very low-level development efforts" required for specific products.

Apple is known for the simplicity, elegance, and ease of use of its products. The company focuses on design as much as it does on function. Jobs stated that great design isn't simply for aesthetic value—it's about function. The process of making products aesthetically pleasing comes from a fundamental desire to make them easy to use. As Jobs once said, "Design is not just what it looks and feels like. Design is how it works."

Big Data in Game Design

Another area of technology in which Big Data plays a key role is in the design of games. Analytics allows game designers to evaluate new retention and monetization opportunities and to deliver more satisfying gaming experiences, even within existing games. Game designers can look at metrics like how much it costs to acquire a player, retention rates, daily active users, monthly active users, revenue per paying player, and session times, that is, the amount of time that players remain engaged each time they play.[7]

Kontagent is one company that provides tools to gather such data. The company has worked with thousands of game studios to help them test and improve the games they create.

Game companies are creating games with completely customizable components. They use a content pipeline approach in which a game engine can import game elements, including graphical elements, levels, objects, and challenges for players to overcome.[8]

---

[7]http://kaleidoscope.kontagent.com/2012/04/26/jogonuat-ceo-on-using-data-driven-game-design-to-acquire-high-value-players/
[8]http://www.cis.cornell.edu/courses/cis3000/2011sp/lectures/12-DataDriven.pdf

The pipeline approach means that game companies can separate different kinds of work—the work of software engineers from that of graphic artists and level designers, for example. It also makes it far easier to extend existing games by adding more levels, without requiring developers to rewrite an entire game.

Instead, designers and graphic artists can simply create scripts for new levels, add new challenges, and create new graphic and sound elements. It also means that not only can game designers themselves add new levels but players can potentially add new levels, or at least new graphical objects.

Separating out the different components of game design also means that game designers can leverage a worldwide workforce. Graphic artists might be located in one place while software engineers are located in another.

Scott Schumaker of Outrage Games suggests that a data-driven approach to game design can reduce the risks typically associated with game creation. Not only are many games never completed, but many completed games are not financially successful. As Schumaker points out, creating a great game isn't[9] just about designing good graphics and levels, it's also about making a game fun and appealing.

It's difficult for game designers to assess these kinds of factors before they implement games, so being able to implement, test, and then tweak a game's design is critical. By separating out game data from the game engine, it becomes far easier to adjust game play elements, such as the speed of ghosts in Pac-Man.

One company that has taken data-driven game design to a new level is Zynga. Well-known for successful Facebook games like CityVille and Mafia Wars, Zynga evaluates the impact of game changes nearly in real-time. Zynga's game makers can see how particular features impact how many gifts people send to each other and whether people are spreading a game virally or not.[10]

By analyzing data, Zynga was able to determine that in FrontierVille it was too hard for new players to complete one of their first tasks, which was building a cabin. By making the task easier, a lot more players ended up sticking around to play the game. Although Zynga's[11] public market value has recently declined, there's clearly a lot to be learned from its approach to game design.

---

[9]http://ai.eecs.umich.edu/soar/Classes/494/talks/Schumaker.pdf
[10]http://www.gamesindustry.biz/articles/2012-08-06-zyngas-high-speed-data-driven-design-vs-console-development
[11]http://www.1up.com/news/defense-zynga-metrics-driven-game-design

Better Cars with Big Data

What about outside the tech world? Ford's Big Data chief John Ginder believes the automotive company is sitting on immense amounts of data that can "benefit consumers, the general public, and Ford itself."[12] As a result of Ford's financial crisis in the mid-2000's and the arrival of new CEO Alan Mulally in 2006,[13] the company has become a lot more open to making decisions based on data, rather than on intuition. The company is considering new approaches based on analytics and simulations.

Ford had analytics groups in its different functional areas, such as for risk analysis in the Ford Credit group, marketing analysis in the marketing group, and fundamental automotive research in the research and development department. Data played a big role in the company's turnaround, as data and analytics were called upon not just to solve tactical issues within individual groups but to be a critical asset in setting the company's go-forward strategy. At the same time, Mulally places a heavy emphasis on a culture of being data-driven; that top-down focus on measurement has had a huge impact on the company's use of data and its turnaround, according to Ginder.

Ford also opened a lab in Silicon Valley to help the company access tech innovation. The company gets data from some four million vehicles that have in-car sensing capabilities. Engineers can analyze data about how people use their cars, the environments they're[14] driving in, and vehicle response.

All of this data has the potential to help the company improve car handling, fuel economy, and vehicle emissions. The company has already used such data to improve car design by reducing interior noise, which was interfering with in-car voice recognition software. Such data also helped Ford engineers determine the optimal position for the microphone used to hear voice commands.[15]

Big Data also helps car designers create better engines. Mazda used tools from MathWorks to develop its line of SKYACTIV engine technologies. Models allow Mazda engineers to "see more of what's going on inside the engine," and achieve better fuel efficiency and engine performance as a result. Such models allow engine designers to test new engine components and designs before creating expensive prototypes.[16,17]

---

[12]http://www.zdnet.com/fords-big-data-chief-sees-massive-possibilities-but-the-tools-need-work-7000000322/
[13]Mulally retired as CEO in June 2014.
[14]http://blogs.wsj.com/cio/2012/06/20/ford-gets-smarter-about-marketing-and-design/
[15]http://blogs.wsj.com/cio/2012/04/25/ratings-upgrade-vindicates-fords-focus-on-tech/
[16]http://www.sae.org/mags/sve/11523/
[17]http://www.ornl.gov/info/ornlreview/v30n3-4/engine.htm

Historically, the challenge has been that internal combustion engines, which power most vehicles, have been incredibly hard to model. That's because they are inherently complex systems. They involve moving fluids, heat transfer, ignition, the formation of pollutants, and in diesel and fuel injection engines, spray dynamics.

Designers are also using data analytics to make decisions about how to improve racecars, decisions that could eventually impact the cars that consumers buy. In one example, the Penske Racing team kept losing races. To figure out why, engineers outfitted the team's[18] racecars with sensors that collected data on more than 20 different variables such as tire temperature and steering. Although the engineers ran analysis on the data for two years, they still couldn't figure out why drivers were losing races.

Data analytics company Event Horizon took the same data but applied a different approach to understanding it. Instead of looking at the raw numbers, they used animated visualizations to represent changes in the racecars. By using these visualizations, they were quickly able to figure out that there was a lag time between when a driver turned the steering wheel and when a car actually turned. This resulted in drivers making lots of small adjustments, all of which together added up to lost time. It's not enough just to have the right data—when it comes to design as well as other aspects of Big Data, being able to see the data in the right way matters a lot.

■ **Note**   Gathering Big Data is the easy part. Interpreting it is the hard part. But if you get it right, you will gain competitive advantages that will vault you to the next level of excellence and company value.

Big Data and Music

Big Data isn't just helping us build better cars and airplanes. It's also helping us design better concert halls. W.C. Sabine, a lecturer at Harvard University, founded the field of architectural acoustics around the turn of the 20th century.[19,20]

In his initial research, Sabine compared the acoustics of the Fogg Lecture Hall, in which listeners found it difficult to hear, with the nearby Sanders Theater, which was considered to have great acoustics. In conjunction with his assistants, Sabine would move materials such as seat cushions from the Sanders

---

[18]http://blogs.cio.com/business-intelligence/16657/novel-encounter-big-data
[19]http://lib.tkk.fi/Dipl/2011/urn100513.pdf
[20]http://en.wikipedia.org/wiki/Wallace_Clement_Sabine

Theater to the Fogg Lecture Hall to determine what impact such materials had on the hall's acoustics. Remarkably, Sabine and his assistants did this work at night, taking careful measurements, and then replacing all the materials by morning so as not to disturb the daytime use of the two halls.

After much study, Sabine defined the reverberation time or "echo effect," which is the number of seconds required for a sound to drop from its starting level by 60 decibels. Sabine figured out that the best halls have reverberation times between 2 and 2.25 seconds. Halls that have reverberation times that are too long are considered too[21] "live," while halls that have reverberation times that are too short are considered too "dry."

The reverberation time is calculated based on two factors: the room volume and the total absorption area, or the amount of absorption surface present. In the case of the Fogg Lecture Hall, where spoken words remained audible for about 5.5 seconds, an additional 12 to 15 words, Sabine was able to reduce the echo effect and improve the acoustics of the hall. Sabine later went on to help design Boston's Symphony Hall.

Since Sabine's time, the field has continued to evolve. Now data analysts can use models to evaluate sound issues with existing halls and to simulate the design of new ones. One innovation has been the introduction of halls that have reconfigurable geometry and materials, which can be adjusted to make a hall optimal for different uses, such as music or speech.

Ironically, classic music halls, such as those built in the late 1800s, have remarkably good acoustics, while many halls built more recently do not. In the past, architects were constrained by the strength and stiffness of timber, which historically was used in the construction of such halls. More recently, the desire to accommodate more seats as well as the introduction of new building materials that have enabled architects to design concert halls of nearly any size and shape has increased the need for data-driven hall design.[22]

Architects are now trying to design newer halls to sound a lot like the halls of Boston and Vienna. Acoustic quality, hall capacity, and hall shape may not be mutually exclusive. By taking advantage of Big Data, architects may be able to deliver the sound quality of old while using the building materials and accommodating the seating requirements of today.

---

[21]http://www.aps.org/publications/apsnews/201101/physicshistory.cfm
[22]http://www.angelfire.com/music2/davidbundler/acoustics.html

Big Data and Architecture

Concert halls aren't the only area of architecture in which designers are employing Big Data. Architects are applying data-driven design to architecture more generally. As Sam Miller of LMN, a 100-person architecture firm points out, the old architectural design model was design, document, build, and repeat. It took years to learn lessons and an architect with 20 years of experience might only have seen a dozen such design cycles.

With data-driven approaches to architecture, architects have replaced this process with an iterative loop: model, simulate, analyze, synthesize, optimize, and repeat. Much as engine designers can use models to simulate engine performance, architects can now use models to simulate the physical act of building.[23]

According to Miller, his group can now run simulations on hundreds of designs in a matter of days and they can figure out which factors have the biggest impact. "Intuition," says Miller, "plays a smaller role in the data-driven design process than it did in the analog process." What's more, the resulting buildings perform measurably better.

Architects don't bill for research and design hours, but Miller says that the use of a data-driven approach has made such time investments worthwhile because it gives his firm a competitive advantage.

Big Data is also helping design greener buildings by putting to work data collected about energy and water savings. Architects, designers, and building managers can now evaluate benchmark data to determine how a particular building compares to other green buildings. The EPA[24]'s Portfolio Manager is one software tool that is enabling this approach. The Portfolio Manager is an interactive energy management tool that allows owners, managers, and investors to track and assess energy and water usage across all of the buildings in a portfolio.[25]

Sefaira offers web-based software that leverages deep physics expertise to provide design analysis, knowledge management, and decision support capabilities. With the company's[26] software, users can measure and optimize the energy, water, carbon, and financial benefits of different design strategies.

---

[23]http://www.metropolismag.com/pov/20120410/advancing-a-data-driven-approach-to-architecture
[24]http://www.greenbiz.com/blog/2012/05/29/data-driven-results-qa-usgbcs-rick-fedrizzi
[25]http://www.energystar.gov/index.cfm?c=evaluate_performance.bus_portfoliomanager
[26]http://venturebeat.com/2012/04/10/data-driven-green-building-design-nets-sefaira-10-8-million/

Data-Driven Design

In studying the design approaches of many different companies and the ways in which data is used, what's clear is that data is being used more and more to inform design. But it is also clear that design, and the disruption that comes from making big changes, still relies on intuition, whether at Apple, Facebook, or at your own company.

As Brent Dykes—evangelist of customer analytics at the web analytics company Adobe/Omniture and author of the blog AnalyticsHero—notes, creative and data are often seen as being at odds with each other. Designers frequently view data as a barrier to creativity, rather than as an enabler of better design.[27]

In one famous instance, Douglas Bowman, a designer at Google, left the company, citing its oppressive, data-driven approach to design. Bowman described an instance in which a team at Google couldn't decide between two shades of blue, so they tested 41 shades between each blue to determine which one performed the best.

Yet Bowman, now Creative Director at Twitter, didn't fault Google for its approach to design, which he describes as reducing "each decision to a simple logic problem,"[28] given the billions of dollars of value at stake. But he did describe data as "a crutch for every decision, paralyzing the company and preventing it from making any design decisions."

In contrast, Dykes believes that the restrictions that data introduces increases creativity. Data can be incredibly helpful in determining whether a design change helps more people accomplish their tasks or results in higher conversions from trial to paid customers on a web site.

Data can help improve an existing design, but it can't, as Facebook designer Adam Mosseri points out, present designers with a completely new design. It can improve a web site, but it can't yet create a whole new site from scratch if that's what is required. Put another way, when it comes to design, data may be more helpful in reaching a local maximum than a global one.

Data can also tell you when a design simply isn't working. As serial entrepreneur and Stanford University lecturer Steve Blank once said to an entrepreneur who was getting his advice, "look at the data." Blank was highlighting that the entrepreneur's thesis simply wasn't bearing out.

---

[27]http://www.analyticshero.com/2012/12/04/data-driven-design-dare-to-wield-the-sword-of-data-part-i/
[28]http://stopdesign.com/archive/2009/03/20/goodbye-google.html

What's also clear across many different areas of design, from games to cars to buildings, is that the process of design itself is changing. The cycle of creating a design and testing it is becoming a lot shorter due to the use of Big Data resources.

The process of getting data on an existing design and figuring out what's wrong or how to incrementally improve it is also happening much faster, both online and offline. Low-cost data collection and computing resources are playing a big role in making the process of design, testing, and redesign a lot faster. That, in turn, is enabling people to have not only their designs, but the design processes themselves, informed by design.

Big Data in Better Web Site Design

As compelling as all that is, however, many of us will never get to design a smartphone, a car, or a building. But creating a web site is something nearly anyone can do. And millions of people do.

As of September 2014, there were more than one billion web sites online.[29] Web sites aren't just the purview of big companies. Small business owners and individuals have built more than 50 million web sites using a free web site design tool called Wix.[30]

While web analytics—the tools used to track site visits, test whether one version of a web page works better than another, and measure conversions of visitors to customers—has come a long way in the last decade, in terms of becoming data-driven, web design itself has not progressed much in the same period of time, says Dykes.

Ironically, the web is one of the easiest forms of design to measure. Every page, button, and graphic can be instrumented. Designers and marketers can evaluate not just on-site design but the impact of advertising, other sites a user has visited, and numerous other off-site factors.

There are lots of web analytics tools available, but many of those tools require heavy analytics expertise or are primarily targeted at technologists, not marketers. Solutions like Adobe Test&Target (formerly Omniture Test&Target) and Google Analytics Content Experiments provide the ability to test different designs, but still require technical expertise. More recently introduced offerings, such as Optimizely, hold the promise of making, creating, and running site optimization tests—known as A/B tests for the way in which they evaluate two variants of the same design—a lot easier.

---

[29]http://www.internetlivestats.com/total-number-of-websites/
[30]http://www.wix.com/blog/2014/06/flaunting-50m-users-with-50-kick-ass-wix-sites/

What's more, at large companies, making changes to a company's web site remains a time-consuming and difficult process, one that involves working with relatively inflexible content management systems and quite often, IT departments that already have too much work to do. Thus, while experimenting with new designs, graphics, layouts, and the like should be easy to do, it's still quite difficult. A web site overhaul is usually a major project, not a simple change.

Many content management systems rely on a single template for an entire site and while they make it easy to add, edit, or remove individual pages, create a blog post, or add a whitepaper, changing the actual site design is hard. Typically there's no integrated, easy way to tweak a layout or test different design approaches. Such changes usually require the involvement of a web developer.

These kinds of systems frequently lack built-in staging capabilities. A major change is either live or it isn't. And unlike platforms like Facebook, which has the ability to roll out changes to a subset of users and get feedback on what's working and what's not, most content management systems have no such capability. Facebook can roll out changes to a subset of users on a very finely targeted basis—to a specific percentage of its user base, males or females only, people with or without certain affiliations, members of certain groups, or based on other characteristics.

This makes changes and new features relatively low risk. In contrast, major changes to most corporate web sites are relatively high risk due to the inability to roll out changes to a subset of a site's visitors. What's more, most content management systems use pages and posts as the fundamental building blocks of design. The other design elements are typically static. In contrast, a site like Facebook may have many modules that make up a particular page or the site as a whole, making any individual module easier to change.

Moreover, marketing executives are often concerned about losing inbound links that reference a particular piece of content. As a result, marketing executives and IT managers alike are loath to make and test such changes on a frequent basis.

The good news is that with the advent of Big Data tools for design and analysis, there are lots of resources available to help marketers understand how to design their sites for better conversion. The lack of data-driven design when it comes to web sites may have more to do with the limitations of yesterday's underlying content management systems and the restrictions they impose than with the point tools available to inform marketers about what they need to fix. As data-driven optimization tools continue to show users the power of being able to tweak, analyze, and tune their sites, more users will demand better capabilities built into their content management systems, which will cause such systems to become more data-driven—or will result in the emergence of new systems that are built with data-driven design in mind from the outset.

Web site designers and marketers recognize that they need to change. Static web sites no longer garner as much traffic in Google. Fresh, current web sites that act a lot more like news sites get indexed better in Google and are more likely to have content shared on social media such as Twitter, Facebook, LinkedIn, blogs, and news sites.

Newer sites, and modern marketing departments, are a lot more focused on creating and delivering content that appeals to actual prospects, rather than just to Google's web site crawlers. In large part, this is due to Google's recent algorithm changes, which have placed more emphasis on useful, current, and authoritative content.

Going forward, it will no longer be enough to have a poor landing page for a customer to land on. Web sites and mobile applications will need to be visually appealing, informative, and specifically designed to draw prospects in and convert them into customers—and then retain them for the long term.

As social media and search change to become more focused on content that's interesting and relevant to human beings, marketers and web site designers will need to change too. They'll need to place more emphasis on data-driven design to create web sites and mobile applications, as well as content that appeals to human beings, not machines. Both on and off the web, we will evolve to create designs and implement design approaches informed by data. Of course, the design tools we use will evolve as well.

In the years ahead, data will continue to become a more integral part of product design, for both digital and physical products. A/B testing will become mainstream as tools for testing different design variants continue to become more automated and easier to use. Expect products to be built in more modular ways so that newer components can easily replace older ones as data shows what works and what doesn't. Continued reduction in the cost of new 3D printing technologies will also make it easier to develop customized products based on what the data indicates.

Why a Picture is Worth a Thousand Words

How Big Data Helps Tell the Story

It's your first visit to Washington, D.C. Arriving in the capital of the United States, you're excited to visit all the monuments and museums, see the White House, and scale the Washington Monument. To get from one place to another, you need to take the local public transit system, the Metro. That seems easy enough. There's just one problem: you don't have a map.[1]

Instead of a map, imagine that the person in the information booth hands you an alphabetized list of stations, train line names, and geographic coordinates. In theory, you have all the information you need to navigate the D.C. metro. But in reality, figuring out which line to take and where to get on and off would be a nightmare.

---

[1] I was searching for a compelling example of why visualizing information is so important when I came across the Washington, D.C. Metro map in the Infographic page on Wikipedia. See http://en.wikipedia.org/wiki/Infographic.

Fortunately, the information booth has another representation of the same data. It's the Washington, D.C. subway map, and it shows all the stations in order on different lines, each in a different color. It also shows where each line intersects so that you can easily figure out where to switch lines. All of a sudden, navigating the Metro is easy.

The subway map doesn't just give you data—it gives you knowledge.

Not only do you know which line to take, but you know roughly how long it'll take to get to your destination. Without much thought, you can see that there are eight stops to your destination, stops that are a few minutes apart each, so it'll take a bit more than 20 minutes to get from where you are to, say, the Air and Space Museum. Not only that but you can recognize each of the lines on the Metro not just by the name or final destination, but by the color as well: red, blue, yellow, green, or orange. Each line has a distinct color that you can recognize on the map—and on the walls of the metro when you're trying to find the right line.

This simple example illustrates the compelling nature of visualization. With a mix of color, layout, markings, and other elements, a visualization can show us in a few seconds what plain numbers or text might take minutes or hours to convey, if we can draw a conclusion from them at all.

To put things in perspective, the Washington, D.C. Metro has a mere 86 stations. The Tokyo subway, which consists of the Tokyo Metro and the Toei, has some 274 stations. Counting all of the railway networks in the greater Tokyo area, there are some 882 stations in total.[2] That number of stations would be virtually impossible to navigate without a map.

Trend Spotting

If you've ever used a spreadsheet, you've experienced first-hand how hard it can be to spot trends in a mass of number-filled cells. Table 5-1 is an example of U.S. Census Data on just the county of Alameda, California from 2010.

---

[2]http://en.wikipedia.org/wiki/Tokyo_subway.

Table 5-1. U.S. Census population data in tabular form

SUMLEV	STATE	COUNTY	STNAME	CTYNAME	YEAR	AGEGRP	TOT_POP	TOT_MALE	TOT_FEMALE
50	6	1	California	Alameda	1	0	1510271	740573	769698
50	6	1	California	Alameda	1	1	97652	50259	47393
50	6	1	California	Alameda	1	2	94546	48145	46401
50	6	1	California	Alameda	1	3	91070	46403	44667
50	6	1	California	Alameda	1	4	100394	51445	48949
50	6	1	California	Alameda	1	5	107049	54515	52534
50	6	1	California	Alameda	1	6	113597	56185	57412
50	6	1	California	Alameda	1	7	114607	56256	58351
50	6	1	California	Alameda	1	8	115275	57062	58213
50	6	1	California	Alameda	1	9	112216	55959	56257
50	6	1	California	Alameda	1	10	114111	56168	57943
50	6	1	California	Alameda	1	11	108506	53431	55075
50	6	1	California	Alameda	1	12	94648	45580	49068
50	6	1	California	Alameda	1	13	78854	37509	41345
50	6	1	California	Alameda	1	14	52663	24568	28095
50	6	1	California	Alameda	1	15	37774	17060	20714
50	6	1	California	Alameda	1	16	29185	12617	16568
50	6	1	California	Alameda	1	17	23391	9144	14247
50	6	1	California	Alameda	1	18	24733	8267	16466

Unlike in the move *The Matrix*, where numbers look like images and images look like numbers, spreadsheets aren't quite as easy to interpret. That's one reason programs like Microsoft Excel and Apple Numbers come with built-in capabilities for creating charts. That census data shown in Table 5-1 is a lot easier to understand when we see it in graphical form, as shown in Figure 5-1.

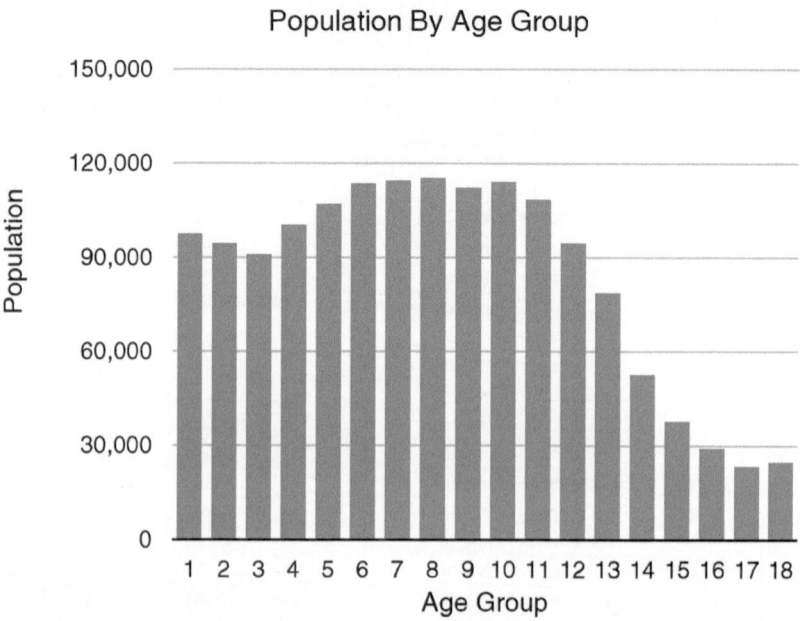

Figure 5-1. U.S. Census population data by age in visual form

When we see a graph like a pie or bar chart, it's often a lot easier to see how things are changing over time or on a relative basis.

How things change over time is critical when making decisions. A single data point, by itself, is often insufficient to tell you how things are going, regardless of whether you're looking at sales trends or health data.

Figure 5-2 shows the U.S. Census Bureau data on new home sales starting in the year 2000. If we were to look just at the value for January 2000, which is 873,000, that wouldn't tell us much by itself. But when we look at new home sales over time, the story is crystal clear. We can see just how dramatic a difference there was between new home sales at the peak of the housing bubble and new homes sales today.[3]

---

[3]Chart generated via www.census.gov. The actual query is https://www.census.gov/econ/currentdata/dbsearch?program=RESSALES&startYear=2000&endYear=2014&categories=ASOLD&dataType=TOTAL&geoLevel=US&adjusted=1&submit=GET+DATA.

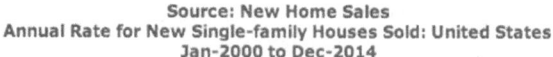

Figure 5-2. U.S. Census new home sales data visualized over time

Using this kind of visual trend analysis is a key way to understand data. Investors, for example, often evaluate a company's performance over time. A company might report revenue and profits for a given quarter. Without a view of financial performance during previous quarters, investors might conclude that the company is doing well.

But what that moment-in-time data can't tell the investors is that the company's sales have been growing less and less each quarter. So while sales and profits in the abstract seem to be good, in reality, the company will be headed for bankruptcy if it doesn't find a way to increase profits.

Internal context is one of the key indicators managers and investors use to figure out how business is trending. Managers and investors also need *external context,* which tells them how they're doing relative to others.

Suppose that sales are down for a given quarter. Managers might conclude that their company isn't executing well. In reality, however, sales might be off due to larger industry issues—for example, fewer homes being built in the case of real estate or less travel, in the case of the airline industry. Without external context, that is, data on how other companies in their industry did over the same time period, managers have very little insight into what's really causing their business to suffer.

Even when managers have both internal and external context, it's still hard for them to tell what's going on just by looking at numbers in the abstract. That's where visualizations can really help.

The Many Types of Visualizations

Nearly every business user is familiar with the well-known pie chart, bar chart, or line graph. These forms of visualization are just the tip of the iceberg when it comes to converting data into its visual equivalent. There are many other types of visualizations as well.

Geographic visualizations are useful for displaying location information. Geographic visualizations often have additional information layered into them. For example, they can show population densities, store locations, income distributions, weather patterns, and other kinds of data that are helpful to see on a visual basis. Figure 5-3 combines geographic information (a map of the United States) with weather data to illustrate just how much of the country is suffering from drought as of August, 2014.[4]

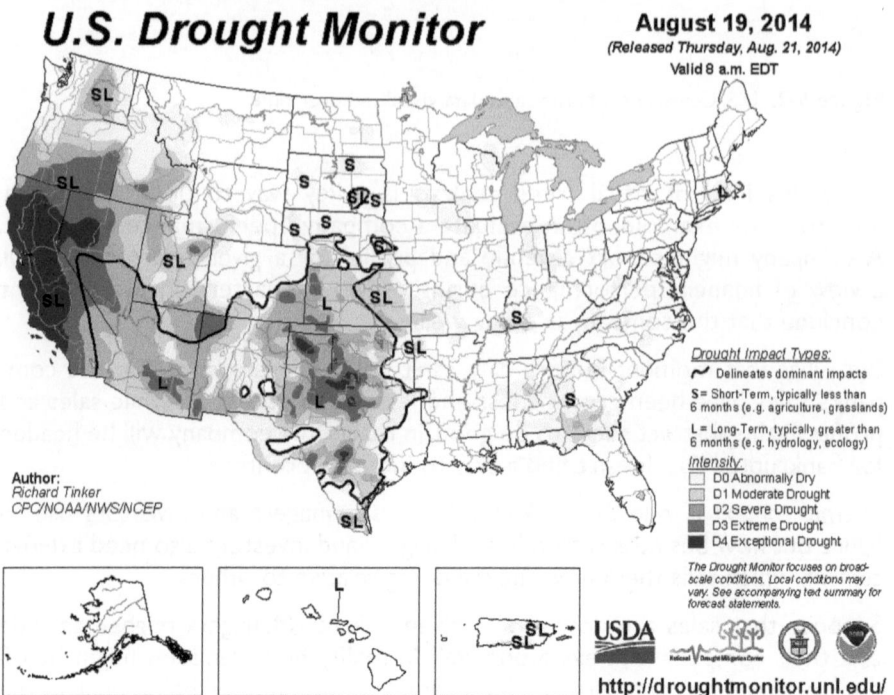

Figure 5-3. A visualization that combines geographic and weather data

---

[4]http://www.ncdc.noaa.gov/news/us-drought-monitor-update-august-5-2014 Produced by the National Drought Mitigation Center at the University of Nebraska-Lincoln, the United States Department of Agriculture, and the National Oceanic and Atmospheric Administration.

Maps can show routing information, telling sales people which locations to visit and when or showing drivers the optimal route to take from one location to another.

Network diagrams show connections and interconnections. Network diagrams can illustrate the way information flows in an organization by showing the relationships between people. Network diagrams can also show connections in a social network or connections between different machines in a computer network.

Time series visualizations illustrate how things change over time. A time series chart might show the consumption of natural resources such as gas, oil, and coal over a period of many years. Or it could show sources of revenue. Time series visualizations can be combined with geographic visualizations to show how the density of populations, or the earning power of certain populations, changes over time.

Infographics are frequently used for marketing purposes, and they don't just show data in visual form but they also incorporate drawings, text, and graphics that tell a story about the data.

Word maps, like the one shown in Figure 5-4, are useful ways to visualize the most frequently mentioned words in large quantities of text.[5] Such visualizations make it easy to determine what a particular body of text is all about. You can create word maps using a variety of tools. One easily accessible web tool is called Wordle, located at `wordle.net`.

Figure 5-4. A word map of the Constitution of the United States

---

[5]Created by pasting the text of the U.S. Constitution into wordle.net.

More and more visualizations are being created that are dynamic in nature. Rather than the static, fixed visualizations of the past, today's interactive visualizations enable you to interact with them so that you can change the time period viewed, zoom in on certain geographic areas for more detail, or change the combinations of variables included in the visualizations to look at the data in a different way. Interactive visualizations combine the best characteristics of traditional visualizations—the power of seeing data presented in graphical form—with access to modern, dynamic analytical capabilities that are easy to use.

■ **Note**   Many sites now showcase the incredible range of visualizations being created on a daily basis. Two such sites are visualizing.org and www.informationisbeautiful.net. The visualizations on these sites can serve as an excellent source of inspiration for creating your own compelling visualizations.

How to Create Visualizations

A number of easy-to-use tools are available to help you create your own visualizations. Visualization tools are available both online and in desktop and mobile versions. Google Public Data Explorer is one great way to get started with creating visualizations. Available at https://www.google.com/public-data/directory, the Public Data Explorer comes loaded with lots of different types of publicly available data. Without installing any software, you can experiment with a variety of different visualizations and view changes in various data sets over time.

There are also online tools available for creating specific types of visualizations. CartoDB (cartodb.com), for example, is a useful tool for creating geographic visualizations. Using CartoDB, it is easy to embed interactive visualizations of complex geographic data sets into your web site, blog, or other application with just a few lines of code.

If you're building your own application, HighCharts (www.highcharts.com) is another visualization resource available online. With very few lines of code, you can load online data into HighCharts and it will do the hard work of displaying that data in chart form.

Due to compliance, privacy, or security requirements, you won't always be able to upload data to a cloud-based visualization tool. In that case, you can use a desktop software application like Tableau Desktop or QlikTech's QlikView. If you need to access data stored in a data repository like Hadoop, Microsoft SQL Server, Oracle, Teradata, or other data sources, you can use Tableau and QlikView to connect directly to these data sources.

These programs can also connect to file-based data sources like Excel files and text files. This means you can access a wide range of different data sources, as well as data sources stored in multiple data repositories, and easily visualize the data contained there.

Software like Tableau Desktop (see Figure 5-5) makes it extremely easy to switch between different kinds of visualizations. That means you can take complex data sets and try out a variety of visualizations quickly to see which one presents your data in the most compelling manner.

For **scatter plots** try
0 or more dimensions
2 to 4 measures

Figure 5-5. The Show Me popup in Tableau Desktop allows users to switch easily between different kinds of visualizations (Courtesy Tableau Software; used with permission)

As you can see in Figure 5-6, with the right tool, it's easy to take otherwise hard to interpret data like a sales forecast and view that in compelling, visual form.

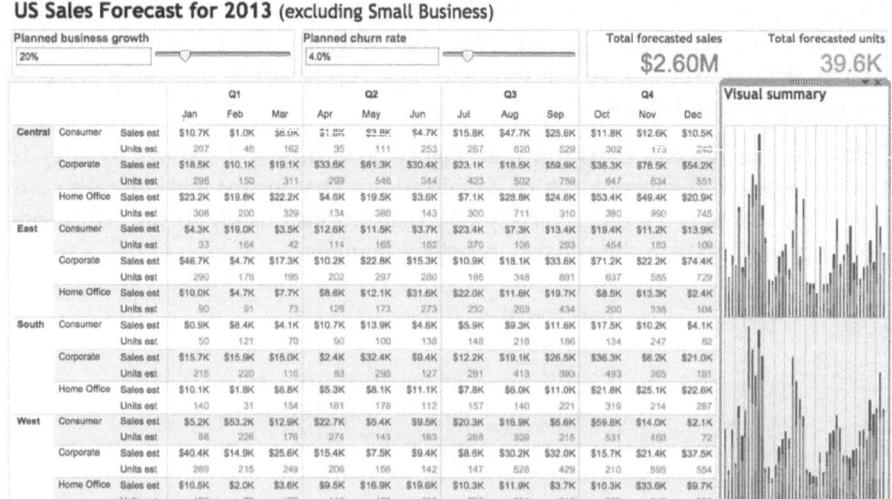

Figure 5-6. Sample sales forecast data shown in Tableau Desktop with a corresponding visualization shown to the right (Courtesy Tableau Software; used with permission)

Regardless of the tool you choose, visualizations make complex data easy to understand. You can not only use visualizations in your presentations, but you can also embed them directly into web sites and applications.

Using Visualization to Compress Knowledge

As the saying goes, a picture is worth a thousand words. But that begs the question of why visualization is so powerful. As visualization expert David McCandless puts it, "visualization is a form of knowledge compression."[6] One form of compression is reducing the size of the data, say by representing a word or a group of words using shorthand, such as a number. But while such compression makes data more efficient to store, it does not make data easier to understand.

A picture, however, can take a large quantity of information and represent it in a form that's easy to understand. In Big Data, such pictures are referred to as visualizations.

Subway maps, pie charts, and bar graphs are all forms of visualization. Although visualization might seem like an easy problem at first, it's hard for a few reasons. First, it's frequently hard to get all the data that people want to visualize

---

[6]https://www.ted.com/talks/david_mccandless_the_beauty_of_data_
visualization/transcript.

into one place and in a consistent format. Internal and external context data might be stored in two different places. Industry data might be in a market research report while actual company sales data may be stored in a corporate database.

Then, the two forms of data might come in slightly different formats. Company sales data might be stored on a daily basis while industry data might be available only on a quarterly basis.

Alternatively, the names given to particular pieces of data might be different; a hard drive might be referred to as "hard drives" in an industry report but referred to by model number in an internal sales database. Such forms of data inconsistency can make it hard to understand what the data is really telling us. There is no silver bullet solution to data inconsistency issues, but newer products like Trifacta and others are emerging to make the problem easier to address.

■ **Tip** There remains an large opportunity to build an easy-to-use hosted data-cleansing service. Today many data consistency issues consume time that data analysts might otherwise apply to solving business problems. A hosted data cleansing and data consistency service could solve this problem on a large scale using a combination of algorithmic and human approaches.

The good news is that modern visualization products can connect directly to a variety of data sources, from local files to databases to data stores like Hadoop. By taking all that data and creating a picture of it, the data can become more than data. It can become knowledge that we can act on.

Visualization is a form of knowledge compression because a seemingly simple image can take vast amounts of structured or unstructured data and compress it into a few lines and colors that communicate the meaning of all that data quickly and efficiently.

Why Is Visual Information So Powerful?

When it comes to visualization, few people have had as big an impact on the field as Edward Tufte. *The New York Times* called Tufte the "Leonardo da Vinci of data."

In 1982, Tufte published one of the defining books of the 20th century, *Visual Display of Quantitative Information*. Although he began his career teaching courses on political science, Tufte's life work has been dedicated to understanding and teaching information design.

One of Tufte's contributions is a focus on making every piece of data in an illustration matter, and excluding any data that isn't relevant. Tufte's images don't just communicate information; many consider his graphics to be works of art. Visualizations are not only useful as business tools, Tufte demonstrates, they can also communicate data in a visually appealing way.

Although it may be difficult to match some of the graphical approaches that Tufte popularized, *infographics*, as they are now commonly known, have become popular ways to communicate information.

Infographics don't just look good. As with other aspects of Big Data, there is a scientific explanation for what makes visual representations of data so compelling.

In a blog post, Tufte cites a press release about an article published in *Current Biology* that describes just how much information we visually absorb.[7] According to the article, researchers at the University of Pennsylvania School of Medicine estimated that the human retina "can transmit visual input at about the same rate as an Ethernet connection."[8]

For their study, the researchers used an intact retina from a guinea pig combined with a device called a multi-electrode array that measured spikes of electrical impulses from ganglion cells. Ganglion cells carry information from the retina to the brain. Based on their research, the scientists were able to estimate how fast all the ganglion cells—about 100,000 in total—in a guinea pig retina transmit information. The scientists were then able to calculate how much data the corresponding cells in a human retina transmitted per second. The human retina contains about one million ganglion cells. Put all those cells together and the human retina transmits information at about 10 megabits per second.

To put that in context, Tor Norretranders, a Danish popular science author, created a graphic illustrating the bandwidth of our senses. In the graphic he showed that we receive more information visually than through any of our other senses. If we receive information via sight at about the same rate as a computer network, we receive information through touch at about one tenth that rate, about the rate that a USB key interfaces with a computer.

We receive information through our ears and nose at an even slower rate, about one tenth of that of touch or about the same speed at which a hard drive interfaces with a computer; and we receive information through our taste buds at a slower rate still.

---

[7]http://www.edwardtufte.com/bboard/q-and-a-fetch-msg?msg_id=0002NC.
[8]http://www.eurekalert.org/pub_releases/2006-07/uops-prc072606.php.

In other words, we get information through our eyes at a rate that is ten to a hundred times faster than through any of our other senses. So it makes sense that information communicated visually is incredibly powerful. And if that information contains a lot of data compressed into a visualization full of knowledge, we can receive that information even faster.

But that's not the only reason such visual data representations are so powerful. The other reason is that we love to share and in particular, we love to share images.

Images and the Power of Sharing

On November 22, 2012, users of photo sharing service Instagram shared a lot of photos. It was Instagram's busiest day ever, with users of the service sharing twice the number of photos on that day as they had the day before. That's because November 22 wasn't just any day of the year. It was Thanksgiving day. Users of Instagram uploaded some 10 million photos that mentioned Thanksgiving-themed words in their captions. To put it mildly, that's a lot of turkey photos, and photos of loved ones too, of course. Some 200 million people now use the service on a monthly basis.[9]

Early in 2012, Facebook purchased Instagram for a billion dollars. Facebook is no slouch either when it comes to sharing photos. Facebook's users were uploading an average of 350 million photos a day as of the end of 2013, more than 10 billion photos every month.[10]

There's another reason we love photos, of course, and that is that they are now so easy to take. Just a few short years ago, we had to make decisions about which photos to take and which not to, at the moment the image was available. If we were almost out of film, we might have saved the last shot for another day. But today, digital cameras, smartphones, and cheap storage have made it possible to capture a nearly unlimited number of digital images. Just about every smartphone now has a camera built in. That means that it's possible not only to take all those photos but to upload and share them easily as well.

Such ease of capturing and sharing images has shown us just how fun and rewarding the activity can be. So it's only natural that when we come across interesting infographics we want to share them too.

---

[9]http://blog.instagram.com/post/80721172292/200m.
[10]http://internet.org/efficiencypaper.

And just as with photos, it's a lot easier to create infographics today than it was in the past. There's also more incentive for companies to create such graphics. In February 2011, search engine giant Google made a change to its algorithms to reward high-quality web sites, particularly "sites with original content and information such as research, in-depth reports, thoughtful analysis, and so on."[11] As a result, marketers at companies realized they needed to do more in order to get their sites ranked—or listed high—in Google search results.

But what is a marketer with limited information to do in order to create a compelling piece of research? Create an infographic. Infographics can take broad sources of data, mesh them together, and tell a compelling story. These can be stories about the web browser wars raging among Internet Explorer, Chrome, Firefox, and Safari or about job creation when it comes to the crowd funding act. Bloggers and journalists looking for compelling graphics to include in their pieces love such graphics because readers love to look at and share them.

The most effective infographics don't just get posted online—they get shared, and they get shared repeatedly. Some of them go viral, getting shared thousands or even millions of times on social networks like Twitter, Facebook, and LinkedIn, and through good old email.

As the demand for the creation of infographics has risen, so too have the number of companies and services available to help create them. One relatively new entrant is Visual.ly, a marketplace for creating infographics. Founded in 2011, the company specializes in enabling its customers to communicate complex data graphically. CartoDB helps people create geographic visualizations. Even users with little or no statistics or analytics background can use the company's web-based design tools. Geographic visualizations are one of the most compelling ways to represent data like population density, store locations, natural resources, and sales routes.

Leading companies such as QlikTech, with its QlikView product, Tableau Software, and TIBCO with Spotfire, provide products that help people create compelling static and interactive visualizations that are used for reporting, analysis and marketing. Meanwhile, the Google Public Data Explorer lets people explore public data sets, such as population growth and per-capita income, online.

---

[11]http://googleblog.blogspot.com/2011/02/finding-more-high-quality-sites-in.html

Putting Public Data Sets to Use

Business users of visualization tools often think of visualization in terms of creating dashboards. Dashboards take data about sales, marketing, and supply chain and turn that data into meaningful charts that management can review easily.

But the power of visualization extends much further. Public data sets refer to data that is publicly available and frequently collected by governments or government-related organizations. The U.S. Census, first taken in 1790, is one such form of data collection.[12] As a result of the Census, there is a vast amount of information available about the U.S. population, including the composition of the population and its geographic distribution.

It's easy to find public data sets online. One helpful resource is data.gov, which links to an enormous variety of public data sources. These sources include National Weather Service data, labor data, U.S. Census data, economic activity data, and a wide range of other data sources. NASA has a number of interesting data sources on its web site at http://data.nasa.gov/, including images of historical interest and images from the Mars Rover project. Such images provide good examples of unstructured data. They can be extremely useful in creating Big Data-scale image management, categorization, and processing systems.

Another interesting public data set is the Common Crawl Corpus, located at http://commoncrawl.org/. The corpus consists of more than 5 billion web pages. The data set is an incredible resource for developing applications that work with large quantities of unstructured text. It's also useful for developing machine-learning algorithms to detect patterns in text, language-processing software, and search products.

Public data sources can be useful in learning how to work with and visualize large quantities of data. They can also be useful when conducting real-world data analysis. As data storyteller Hans Rosling illustrates, publicly available population and health data is extremely valuable in understanding population changes, the rise and fall of nations, and the progress (or lack thereof) in fighting infant mortality and other epidemics.[13] Rosling uses data visualizations to tell stories with data, in particular public data, in much the way a football commentator uses football replays.

---

[12]http://www.census.gov/history/www/census_then_now/
[13]http://www.ted.com/talks/hans_rosling_the_good_news_of_the_decade.html

Rosling animates data. He doesn't make cartoons out of it. Rather, he plots data on a graph and then shows how that data changes over time—how the relative populations or incomes of different nations evolve over periods of 40 or 50 years, for example. Such animations bring data to life, and the software that Rosling developed with his son and daughter-in-law became the basis for the Google Public Data Explorer.

Some of the most famous visualizations of all time are based on presenting publicly available data in new and compelling ways. Visual.ly showcases a few such charts on its web site in a post entitled *12 Great Visualizations That Made History*.[14] Some of these visualizations illustrate just how impactful the right graphic can be. In one such instance, John Snow's map of Cholera outbreaks in London in 1854 helped explain that water in contaminated wells was responsible for the spread of the disease.

Another famous life-saving chart from around the same time is from Florence Nightingale, known as the mother of modern nursing. Nightingale used a coxcomb diagram to "convey complex statistical information dramatically to a broad audience."[15] See Figure 5-7.

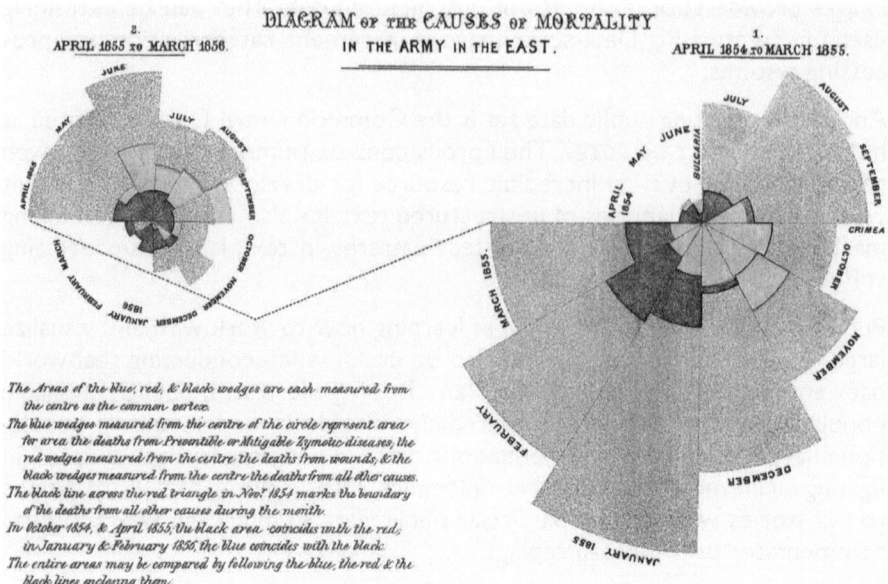

Figure 5-7. Nightingale's diagram of the causes of mortality in the British Army

[14]http://blog.visual.ly/12-great-visualizations-that-made-history/
[15]http://www.datavis.ca/gallery/historical.php

In particular, Nightingale's charts showed that for the British Army, many deaths were preventable. More soldiers died from non-battle causes than battle-related causes. As a result, she was able to convince the government of the importance of using sanitation to decrease mortality rates.

Real-Time Visualization

The information most infographics provide is static in nature, and even the animations Rosling created, compelling as they are, are comprised of historical data.

Frequently, infographics take a long time and a lot of hard work to create: they require data, an interesting story to tell, and a graphics designer who can present the data in a compelling way. The work doesn't stop there; once the graphic is created, like the tree falling in the woods with no one to hear it, the graphic has real value only if it's distributed, promoted, shared, and viewed. By then of course the data itself may be weeks or months old. So what about presenting compelling visualizations of data that are real-time in nature?

For data to be valuable in real time, three things must happen. The data itself must be available, there must be sufficient storage and computer processing power to store and analyze the data, and there must be a compelling way to visualize the data that doesn't require days or weeks of work.

If the idea of knowing what millions of people think about something in real time and being able to illustrate what they think visually seems far-fetched, think again. We need look no further than the 2012 presidential election to be convinced.

In decades past, polling was performed by individual pollsters calling people to ask their opinions or talking to them in person. By combining polls of a relatively small number of people with statistical sampling methods, pollsters were able to make predictions about the outcome of elections and draw conclusions about how people felt about important political issues.

Nielsen used similar forms of statistics for television measurement and comScore did the same for the web. Nielsen originally performed media measurement by using a device to detect to which stations 1,000 people had tuned their radios.[16] The company later applied a related approach to television shows in what became widely known as the Nielsen ratings.

Such forms of measurement are still widely used but as in other areas, Big Data is transforming the way we measure. If there is one company in the last few years that has had more impact on our ability to measure public opinion than any other—an activity known as *sentiment analysis*—it is Twitter.

---

[16]http://en.wikipedia.org/wiki/Nielsen_Company

In fact, Twitter may be one of the most under-appreciated companies around in terms of its Big Data assets. As of October, 2012, Twitter users were sending some 500 million tweets—short text messages—across the network per day,[17] a remarkable amount of human-generated information. That's up from no tweets sent in 2006.

By evaluating the words used in tweets, computer programs can not only detect which topics are *trending*, that is, receiving more attention, but also can draw conclusions about how people feel and what opinions they hold.

Capturing and storing such data is just one aspect of the kind of Big Data challenge and opportunity a company like Twitter faces. To make it possible to analyze such data, the company has to provide access to the stream of tweets, nearly 5,000 text messages per second, and even more during events like presidential debates, when users create some 20,000 tweets per second. Then comes the task of analyzing those tweets for common words, and finally, presenting all that data in a meaningful way. That means converting all those tweets into a visualization.

Handling such massive, real-time streams of data is difficult, but not impossible. Twitter itself provides programmatic interfaces to what is commonly known as the fire hose of tweets. Around Twitter, companies like DataSift have emerged to provide access as well.

Other companies, such as BrightContext, which was recently acquired by WealthEngine, provide tools for real-time sentiment analysis. During the 2012 presidential debates, the *Washington Post* used BrightContext's real-time sentiment module to measure and chart sentiment while viewers watched the debates.[18] Topsy, a real-time search company recently acquired by Apple, has indexed some 200 billion tweets. The company powered Twitter's political index, known as the Twindex. Vizzuality specializes in mapping geospatial data and it powered *The Wall Street Journal*'s election maps.

All of these systems work by processing massive quantities of unstructured data, in the form of text, and presenting that data in visual form.

In contrast to phone-based polling, which is time-consuming and typically costs around $20 per interview, real-time measurement simply costs compute cycles and can be done on an unprecedented scale. Products like those from some of the companies mentioned here can then provide real-time visualizations of the collected data.

---

[17]http://news.cnet.com/8301-1023_3-57541566-93/report-twitter-hits-half-a-billion-tweets-a-day/.
[18]http://www.forbes.com/sites/davefeinleib/2012/10/22/not-your-grandmothers-presidential-debate/5/.

But such visualization doesn't stop at displaying real-time information in web sites. Google Glass,[19] which *Time* magazine called one of the best inventions of 2012, is "a computer built into the frame of a pair of glasses, and it's the device that will make augmented reality part of our daily lives."[20] In the future, not only will we be able to see visual representations of data on our computers and mobile phones, we'll also be able to visualize and understand the physical world better as we move around it.

If that sounds like something out of a science fiction book, it's not. Today, Google Glass costs $1,500 and is somewhat bulky. But just as other new technologies have gotten smaller and cheaper over time, so too will Google Glass. Augmented visualization may very well become part of our daily lives.

Why Understanding Images Is Easy for Us and Hard for Computers

Ironically, while computers excel at processing large amounts of textual information, they still struggle with analyzing visual information. Just recall the last time you took a few hundred photos and wished you had a web site or a piece of software that would automatically weed out the bad photos and group related photos together. Or what about automatically figuring out who is in the photos and sharing copies of those photos with them?

On a larger scale, companies like Facebook have to filter out inappropriate images. Amazon has to determine which textual product descriptions match their image counterparts and which ones don't. These would seem to be relatively easy problems for computers to solve and while the science of image recognition and characterization has advanced significantly, performing such analysis on a large scale remains challenging. Today, human beings still perform many of these recognition and matching tasks.

In their paper, *Why is Real-World Visual Object Recognition Hard?*,[21] scientists from MIT and Harvard stated, "the ease with which we recognize visual objects belies the computational difficulty of this feat. At the core of this challenge is image variation—any given object can cast an infinite number of different images onto the retina, depending on an object's position, size, orientation, pose, lighting, etc."

---

[19]http://www.forbes.com/sites/davefeinleib/2012/10/17/3-big-data-insights-from-the-grandfather-of-google-glass/
[20]http://techland.time.com/2012/11/01/best-inventions-of-the-year-2012/slide/google-glass/
[21]http://www.ploscompbiol.org/article/info:doi/10.1371/journal.pcbi.0040027

Simply put, images can have a lot of variability, making it difficult to tell when different images contain the same objects or people. What's more, pattern detection is more difficult; while the word "president" is easy to find in a sentence and hence relatively easy to find in millions of sentences, it's much harder to recognize the person holding that title in images.

Having an individual human being characterize images is one thing. But what about trying to do it with millions of images? To solve their image-characterization problems, companies like Amazon and Facebook turn to crowdsourcing marketplaces like oDesk and Amazon Mechanical Turk.[22] On these market-places, content moderators who pass certain tests or meet certain qualifications gain access to images and can then characterize and filter them.

Today, computers are good at helping us create visualizations. But tomorrow, as products like Google Glass continue to evolve, they may also help us better understand visual information in real time.

The Psychology and Physiology of Visualization

One industry that understands the importance of presenting information visually better than almost any other is the advertising industry. It is one of several that is on the leading edge of taking advantage of new Big Data technologies.

If there is any doubt that images are a powerful means of communication, we need look no further than the $70 billion that U.S. companies spend each year on TV advertising.[23] As Nigel Hollis, chief global analyst at the market research firm Millward Brown, points out, companies wouldn't spend so much on TV advertising if it didn't work.[24]

Where people get confused about the impact of TV advertising, Hollis says, is in thinking that advertisers want to get them to do something immediately. That's where "they're wrong." Brand advertising doesn't succeed through calls to action or arguments, but rather through leaving positive impressions. "The best advertisements use images, jingles, and stories to focus attention on the brand." In particular, says Hollis, "engaging and memorable ads slip ideas past our defenses and seed memories that influence our behavior."

---

[22]http://gawker.com/5885714/.

[23]http://www.theatlantic.com/business/archive/2011/08/why-good-advertising-works-even-when-you-think-it-doesnt/244252/.

[24]Post on *The Atlantic* entitled "Why Good Advertising Works (Even When You Think It Doesn't)." http://www.theatlantic.com/business/archive/2011/08/why-good-advertising-works-even-when-you-think-it-doesnt/244252/.

In fact, some advertisers have taken the delivery of visual images one step further, applying data analysis to determine which visualizations are most effective through a science called *neuromarketing*. Neuromarketing uses functional magnetic resonance imaging (fMRI) and other technologies "to observe which areas of the brain 'light up'"[25] in response to a variety of advertising approaches. Marketers can even simulate situations to determine which placement, such as on billboards or on the sides of buses, produces the most impact.

Thus visualization is not only an effective way to communicate large quantities of data, it also ties directly into the brain, triggering emotional and chemical responses. Visualization may be one of the best ways to communicate a data-based message. What studies have shown is that it is not just the visualization itself that matters, but when, where, and how such visualizations are presented.

By setting the right context, choosing the right colors, and even selecting the right time of day, it's possible to communicate the insights locked up in vast amounts of data a lot more effectively. As the famous media researcher Marshall McLuhan once said, "the medium is the message." Now scientific evidence is showing just how important context and delivery is when communicating information.

The Visualization Multiplier Effect

As we've seen in this chapter, visualization and data go hand in hand. There are instances, of course, when computers can act on data with no human involvement. For example, it simply wouldn't be possible for humans to figure out which text ads to display alongside search results when dealing with the billions of search queries we make on sites like Google and Bing. Similarly, computer systems excel at automated pricing decisions and evaluating millions of transactions quickly to determine which ones are fraudulent.

But there remain any number of situations in which humans are trying to make better decisions based on data. Just because we have more data available does not mean that it's easier to produce better insights from that data. In fact, the opposite may be true. The more data we have, the more important it becomes to be able to distill that data into meaningful insights that we can act on. Visualizing such data is one of the most powerful mechanisms we have for doing so.

---

[25]http://www.businessweek.com/stories/2007-10-08/this-is-your-brain-on-advertisingbusinessweek-business-news-stock-market-and-financial-advice.

Visualization is effective because our eyes have ultra high throughput into our brains, as much as a hundred times greater throughput than some of our other senses. Visualization can trigger emotional responses. It can compress vast amounts of data into knowledge we can use.

Combine the knowledge compression of visualization with the high throughput of visual delivery and you get the visualization multiplier effect—more data absorbed faster.

Big Data isn't just about the data itself but about how we communicate it and what we do with it. Tools like visualization also mean that Big Data isn't just the domain of scientists, data analysts, or engineers. Big Data, in the form of visualization, is everywhere around us, from the charts we use to make critical business decisions to the advertisements we create to communicate our messages more effectively.

Social media platforms are changing the way we communicate and are enabling the broader distribution not just of textual information but of high-impact visual knowledge. With the right visualization, data is more than just text or numbers. The right visualization can tell a story that has a very real impact not just in business but in broader contexts such as education and global health as well. As you'll see in the next chapter, visualization is just one of the many areas in which Big Data is creating exciting new opportunities.

The Intersection of Big Data, Mobile, and Cloud Computing

Boon to App Development

The intersection of Big Data, mobile, and cloud computing has created the perfect storm for the development of innovative new applications. Mobile visualization applications are providing dynamic insight into sales, marketing, and financial data from virtually anywhere. Cloud computing services are making it possible to store virtually unlimited amounts of data and access scalable, low-cost data processing functionality at the push of a button.

This chapter covers the latest advances in Big Data, cloud computing, and mobile technologies. From choosing the right technologies for your application to putting together your team, you'll learn everything you need to know to build scalable Big Data Applications (BDAs) that solve real business problems.

How to Get Started with Big Data

Building your own BDA need not be hard. Historically, the investment, expertise, and time required to get a new Big Data project up and running prevented many companies and entrepreneurs from getting into the Big Data space. Now, however, Big Data is accessible to nearly anyone, and building your own BDA is well within reach.

Mobile devices mean the insights from Big Data are readily available, while cloud services make it possible to analyze large quantities of data quickly and at a much lower cost than was traditionally possible. Let's first take a look at the latest cloud-based approaches to working with Big Data.

The Latest Cloud-Based Big Data Technologies

Building Big Data applications traditionally meant keeping data in expensive, local data warehouses and installing and maintaining complex software. Two major developments mean that this is no longer the case.

First, the widespread availability of high-speed broadband means that it is easier to move data from one place to another. No longer must data produced locally be analyzed locally. It can be moved to the cloud for analysis.

Second, more and more of today's applications are cloud-based. That means more data is being produced and stored in the cloud. Increasing numbers of entrepreneurs are building new BDAs to help companies analyze cloud-based data such as e-commerce transactions and web application performance data.

The latest cloud-based Big Data technologies have evolved at three distinct layers—infrastructure, platform, and application.

Infrastructure

At the lowest level, Infrastructure as a Service (IaaS), such as Amazon Elastic Compute Cloud (EC2) and Simple Storage Service (S3), as well as Google Cloud Storage, make it easy to store data in the cloud.

Cloud-based infrastructure services enable immense scalability without the corresponding up-front investments in storage and computing infrastructure that is normally required. Perhaps more than any other company, Amazon pioneered the public cloud space with its Amazon Web Services (AWS) offering, which you'll read about in more detail in the next section.

Providers such as AT&T, Google, IBM, Microsoft, and Rackspace have continued to expand their cloud infrastructure offerings. IBM has recently become more aggressive, perhaps growing concerned about the rapid growth of AWS.

In 2013, IBM acquired SoftLayer Technologies for $2 billion to expand its cloud services offering. SoftLayer was doing an estimated $800M to $1B in revenue annually prior to the acquisition.[1]

Amazon, Google, Microsoft and Rackspace have been competing particularly aggressively at the infrastructure level, repeatedly cutting prices on both computing and storage services. Between them, the four companies cut prices some 26 times in 2013, according to cloud management firm RightScale, with Amazon making the most price cuts. This bodes well for Big Data because it means the cost of storing and analyzing large quantities of data in the cloud are continuing to decrease. In many cases, working with Big Data in the cloud is cheaper than doing so locally.

Platform

At the middle layer, Platform as a Service (PaaS) solutions offer a combination of storage and computing services while providing more application-specific capabilities. Such capabilities mean that application developers can spend less time worrying about the low-level details of how to scale underlying infrastructure and more time focusing on the unique aspects of their applications.

After getting off to a somewhat late start, Microsoft has continued to expand its Microsoft Azure cloud offering. Microsoft Azure HDInsight is an Apache Hadoop offering in the cloud, which enables Big Data users to spin Hadoop clusters up and down on demand.

Google offers the Google Compute Engine and Google App Engine, in addition to Google Fusion, a cloud-based storage, analysis, and presentation solution for rapidly working with immense quantities of table-based data in the cloud. More recently, Google introduced Google Cloud DataFlow, which Google is positioning as a successor to the MapReduce approach. Unlike MapReduce, which takes a batch-based approach to processing data, Cloud DataFlow can handle both batch and streaming data.

Newer entrants like Qubole—founded by former Big Data leaders at Facebook—and VMWare—with its CloudFoundry offering—combine the best of existing cloud-based infrastructure services with more advanced data and storage management capabilities. Salesforce.com is also important to mention, since the company pioneered the cloud-based application space back in 1999. The company offers its Force.com platform. While it is well-suited for applications related to customer relationship management (CRM), the platform does not yet offer capabilities specifically designed for BDAs.

---

[1]http://www.forbes.com/sites/bruceupbin/2013/06/04/ibm-buys-privately-held-softlayer-for-2-billion/

Application

Software as a Service (SaaS) BDAs exist at the highest level of the cloud stack. They are ready to use out of the box, with no complex, expensive infrastructure to set up or software to install and manage. In this area, Salesforce.com has expanded its offerings through a series of acquisitions including Radian6 and Buddy Media. It now offers cloud-based social, data, and analytics applications.

Newer entrants like AppDynamics, BloomReach, Content Analytics, New Relic, and Rocket Fuel all deal with large quantities of cloud-based data. Both AppDynamics and New Relic take data from cloud-based applications and provide insights to improve application performance. BloomReach and Content Analytics use Big Data to improve search discoverability for e-commerce sites. Rocket Fuel uses Big Data to optimize the ads it shows to consumers.

The application layer is likely to see continued growth over the next few years as companies seek to make Big Data accessible to an ever broader audience. For end-users, cloud-based BDAs provide a powerful way to reap the benefits of Big Data for specific verticals or business areas without incurring the setup time and costs traditionally required when starting from scratch.

■ **Tip** Employing cloud-based applications is an economical and powerful way to reap the benefits of Big Data intake and analysis at the business and end-user level.

But with many areas not yet addressed by today's BDAs, there are plenty of technologies to help you get started building your own BDA fast. In that regard, perhaps no offering is better known than AWS.

Amazon Web Services (AWS)

Almost no cloud-based platform has had more success than AWS. Until recently, common wisdom was that data that originated locally would be analyzed locally, using on-site computer infrastructure, while data that originated in the cloud would be stored and analyzed there. This was in large part due to the time-consuming nature of moving massive amounts of data from the on-site infrastructure to the cloud so that it could be analyzed.

But all that is changing. The availability of very high bandwidth connections to the cloud and the ease of scaling computing resources up and down in the cloud means that more and more Big Data applications are moving to the cloud or at a minimum making use of the cloud when on-site systems are at capacity.

But before you explore some of the ways in which businesses are using Amazon Web Services for their Big Data needs, first take a step back and look at how AWS became so popular. Although it seems like AWS became a juggernaut virtually overnight, Amazon actually introduced the service in 2003. By 2006, Amazon had commercialized the service and was on its way to becoming the powerhouse that it is today. According to some analyst estimates, AWS accounted for some $3.8 billion of Amazon's revenue in 2013 and is expected to account for a whopping $6.2 billion in 2014.[2]

How did an online bookseller turn into one of the leading providers of public cloud services? Amazon had developed an infrastructure that allowed it to scale virtual servers and storage resources up and down on demand, a critical capability to address the significant spikes in demand the company saw during holiday shopping seasons. Amazon made the storage and computing capacity—quite often excess capacity—that it had originally built to support the demands consumers were placing on its retail web site available to business users.

Of course, cloud-based infrastructure had existed previously in the form of web hosting from providers like GoDaddy, Verio, and others. But Amazon made it truly easy for users to get started by signing up for an account online using a credit card. Amazon made it equally easy to add more storage (known as S3 for Simple Storage Service) and computing power (known as EC2 for Elastic Compute Cloud).

Notably, Amazon's offering provides storage and computing in a utility style model. Users pay only for what they actually consume. That makes it possible to buy a few hours of processing when necessary, rather than committing to a months- or years-long purchase of an entire server. Users can try one server, known as an instance, run it for however long they need, and turn it off when they don't need it anymore. They can bring up bigger or smaller instances based on their computing needs. And all this can be done in a few minutes if not seconds.

Amazon's no-lock-in offering combined with the ease of getting started really sets it apart from the hosting providers that came before it. What's more, users figure that if the infrastructure is good enough to support Amazon's huge e-commerce web site, it is more than good enough to support their own applications.

---

[2]http://www.informationweek.com/cloud/infrastructure-as-a-service/amazons-cloud-revenues-examined/d/d-id/1108058?

And Amazon's pricing is hard to beat. Because of the scale of its infrastructure, Amazon buys immense amounts of hardware and bandwidth. As a result, it has tremendous purchasing power, which in turn enabled it to lower its prices repeatedly. Not only do AWS users get to leverage the capabilities of a flexible cloud-based infrastructure, they also get to take advantage of Amazon's purchasing power as if it were their own. Plus, they don't have to worry about all of the infrastructure requirements that typically go along with maintaining physical servers: power, cooling, network capacity, physical security, and handling hard-drive and system failures. With AWS, all of that comes with the service.

As a result of this flexibility, AWS has seen tremendous adoption. Leading tech companies like Netflix and Dropbox run on top of AWS. But AWS adoption doesn't stop there. Pharmaceutical maker Pfizer uses AWS when its need for high-performance computing (HPC) services outstrips the capacity the company has available on-site. This is a great example of cloud-based Big Data services in action because the HPC workloads used for drug discovery and analysis are often spiky in nature. From time to time, immense amounts of computing power are required for pharmaceutical analytics; the cloud is an ideal way to support such spikes in demand without requiring a long-term purchase of computing resources.

In another example, entertainment company Ticketmaster uses AWS for ticket pricing optimization. Ticketmaster needed to be able to adjust ticket costs but did not want to make the costly up-front investment typically required to support such an application. In addition, as with Pfizer, Ticketmaster has highly variable demand. Using AWS, the company was able to deploy a cloud-based MapReduce and storage capability. According to Amazon, taking a cloud-based approach reduced infrastructure management time by three hours a day and reduced costs by some 80 percent.

Over time, Amazon has added more and more services to its cloud offering, especially services that are useful for BDAs. Amazon Elastic MapReduce (EMR) is a cloud-based service for running MapReduce jobs in the cloud.

MapReduce is a well-known Big Data approach for processing large data sets into results, with the data typically segmented into chunks that are distributed over multiple servers or instances for processing, with the results then put back together to produce the final insights. MapReduce was originally designed by Google in 2004. Apache has popularized the approach with its open source Hadoop MapReduce offering and companies like Cloudera, HortonWorks, and MapR have made the approach even more viable with commercial offerings. Hadoop and the MapReduce approach have become the mainstay of Big Data due to their ability to store, distribute, and process vast amounts of data.

With EMR, users do not need to perform the often laborious and time-consuming setup tasks required to create a MapReduce infrastructure. Moreover, if additional compute resources are needed to complete MapReduce jobs in a timely fashion, users can simply spin up more instances. Perhaps most importantly, users don't need to commit to a huge upfront investment to get their very own MapReduce infrastructure. Instead of buying the hardware and network infrastructure needed for a typical 50- or 100-node computing cluster, users can simply spin up as many instances as they need and then turn them off when their analysis is done.

Amazon's solution doesn't stop there. In the last few years the company has also introduced an ultra low-cost data archiving solution called Glacier. While Glacier does not enable access to data in real-time, it is one-tenth the price or less of traditional archiving solutions. This gives those on the fence yet another reason to move their data to the cloud.

To further expand its offerings, Amazon introduced Kinesis, a managed cloud service for processing of very large quantities of streaming data. Unlike EMR and MapReduce, which are batch-based, Kinesis can work with streaming data such as tweet streams, application events, inventory data, and even stock trading data in real time.

■ **Note** Amazon is continually introducing new Big Data and cloud-related capabilities in Amazon Web Services. Kinesis is the company's latest offering, designed to handle large quantities of streaming data like tweets, stock quotes, and application events.

No longer is it the case that only data that originated in the cloud should be analyzed in the cloud. Now and in the years ahead, low-cost storage, flexible, scalable computing power, and pre-built services like EMR and Kinesis combined with high bandwidth access to public cloud infrastructure like AWS will provide more compelling reasons to perform Big Data analysis in the cloud.

AWS started out with limited customer support and a lack of enterprise-grade Service Level Availability (SLA) options. Now the company provides enterprise-grade availability. Companies like Netflix, Dropbox, Pfizer, Ticketmaster, and others rely on AWS to run their mission-critical applications. Some analysts project Amazon will go from several billion in revenue for AWS to $15 to $20 billion in revenue annually in just the next few years. To do so, it's highly likely the company will continue to expand its enterprise cloud and Big Data offerings.

AWS provides a compelling alternative to traditional infrastructure, enables the rapid launch of new Big Data Services, and provides on-demand scalability. It's also a good option for overflow capacity when it comes to supporting the demands of large-scale analytics applications. What we have seen so far in terms of services offered and near continuous price reductions is just the beginning. There are many more cloud-based Big Data services yet to come, with all of the inherent affordability and scalability benefits.

Public and Private Cloud Approaches to Big Data

Cloud services come in multiple forms—public, private, and hybrid. Private clouds provide the same kind of on-demand scalability that public clouds provide, but they are designed to be used by one organization. Private clouds essentially cordon off infrastructure so that data stored on that infrastructure is separate from data belonging to other organizations.

Organizations can run their own private cloud internally on their own physical hardware and software. They can also use private clouds that are deployed on top of major cloud services such as AWS. This combination often offers the best of both worlds—the flexibility, scalability, and reduced up-front investment of the cloud combined with the security that organizations require to protect their most sensitive data.

■ **Note**   Running a private cloud on top of a major service like AWS provides flexibility, low cost, and all the benefits of public clouds but adds the security that many of today's businesses demand.

Some organizations choose to deploy a combination of private cloud services on their own in-house hardware and software while using public cloud services to meet spikes in demand. This hybrid cloud approach is particularly well-suited for Big Data applications. Organizations can have the control and peace of mind that comes from running their core analytics applications on their own infrastructure while knowing that additional capacity will immediately be available from a public cloud provider when needed.

The compute-intensive parts of many BDAs such as pricing optimization, route planning, and pharmaceutical research do not need to be run all the time. Rather, they require an immense amount of computing power for a relatively short amount of time in order to produce results from large quantities of data. Thus, these applications are perfectly suited to a hybrid cloud approach.

What's more, such applications can often take advantage of the lowest available pricing for computing power. In the case of AWS, spot instances are priced based on market supply and demand. The spot instance buyer offers a price per hour and if the market can support that price, the buyer receives the instance. Spot instance prices are typically much lower than the equivalent prices for on-demand instances.

The downside of spot instances is that they can go away without notice if market prices for instances go up. This can happen if the cloud services provider sees a significant increase in demand, such as during a holiday period. But the transient nature of Big Data processing combined with the distributed nature of the MapReduce approach means that any results lost due to an instance going away can simply be re-generated on another instance.

The more instances over which Big Data processing is distributed the less data is at stake on any one given instance. By combining a distributed architectural approach with the low-cost nature of spot instances, companies can often significantly reduce their Big Data related costs. By layering a virtual private cloud on top of public cloud services, organizations can get the security they need for their Big Data Applications at the same time.

How Big Data Is Changing Mobile

Perhaps no area is as interesting as the intersection of Big Data and mobile. Mobile means that Big Data is accessible anytime, anywhere. What used to require software installed on a desktop computer can now be accessed via a tablet or a smartphone. What's more, Big Data applications that used to require expensive custom hardware and complex software are now available. You can monitor your heart rate, sleep patterns, and even your EKG using a smartphone, some add-on hardware, and an easy-to-install mobile application. You can combine this with the cloud to upload and share all of the data you collect.

Big Data and mobile are coming together not only for the display and analysis of data but also for the collection of data. In a well-known technological movement called the Internet of Things (IoT), smartphones and other low-cost devices are rapidly expanding the number of sensors available for collecting data. Even data such as traffic information, which used to require complex, expensive sensor networks, can now be gathered and displayed using a smartphone and a mobile app.

Fitness is another area where Big Data is changing mobile. Whereas we used to track our fitness activities using desktop software applications, if we tracked them at all, now wearable devices like Fitbit are making it easy to track how many steps we've walked in a given day. Combined with mobile applications and analytics, we can now be alerted if we haven't been active enough. We can even virtually compete with others using applications like Strava combined with low-cost GPS devices. Strava identifies popular segments of cycling routes and lets individual cyclists compare their performance on those segments. This enables riders to have a social riding experience with their friends and even well-known professional cyclists even when they can't ride together. These riders can then analyze their performance and use data to improve.

At the doctor's office, applications like PracticeFusion are enabling doctors to gather and view patient data for Electronic Health Records (EHR) and to analyze that data right on a tablet that they can easily carry with them. No longer does data collection have to be a multiple-step process of writing down notes or recording them as audio files, and then transferring or transcribing those notes into digital form. Now, doctors can record and view patient information "in the moment" and compare such information to historical patient data or to benchmark data. No more waiting—the intersection of Big Data and mobile means real time is a reality.

But the applications for real-time, mobile Big Data don't stop there. Google Glass, a futuristic-looking wearable device that both records and presents information, is changing the way we view the world. No longer do you need to refer to your smartphone to get more information about a product or see where you are on a map. Glass integrates real-time information, visualization, and cloud services, all available as an enhancement to how you see the world.

In the business world, Big Data is changing mobile when it comes to applications like fleet routing, package delivery, sales, marketing, and operations. Analytics applications specifically optimized for mobile devices make it easy to evaluate operational performance while on the go. Financial data, web site performance data, and supply chain data, among many other sources, can all easily be accessed and visualized from tablets and smartphones. Not only do these devices have extremely capable graphics capabilities, but their touch-screen interfaces make it easy to interact with visualizations. Compute-intensive data processing can be done in the cloud, with the results displayed anywhere, anytime.

Big Data is having a big impact in the fields of logistics and transportation as well. Airplane engine data is downloaded in real time to monitor performance. Major automotive manufacturers such as Ford have created connected car initiatives to make car data readily accessible. As a result, developers will be able to build new mobile applications that leverage car data.

Such car applications, once the domain of only the most advanced car racing teams, are already becoming widely available. Opening up access to the wealth of information produced by automobiles means that car manufacturers don't have to be the only producers of interesting new car applications. Entrepreneurs with ideas for compelling car-related applications will be able to create those applications directly.

This platform approach is not limited to cars of course. It makes sense for virtually any organization that has a rich source of data it wants to open up so that others can leverage it to create compelling new applications.

Fitness, healthcare, sales, and logistics are just a few of the many areas where mobile, cloud, and Big Data are coming together to produce a variety of compelling new applications. Big Data is changing mobile—and mobile is changing Big Data.

How to Build Your Own Big Data Applications

You've taken a look at the different kinds of applications you can build with Big Data and the infrastructure available to support your endeavor. Now you'll dig into the details of building your own BDA, including how to identify your business need.

Building BDAs need not be hard. Historically, the investment, expertise, and time required to get a new Big Data project up and running prevented many companies and entrepreneurs from getting into the Big Data space. Now, however, Big Data is accessible to nearly anyone, and building a BDA is well within reach.

To build your own BDA, you need five things:

- A business need

- One or more sources of data

- The right technologies, for storage, computing, and visualization

- A compelling way to present insights to your users

- An engineering team that can work with large-scale data

Defining the Business Need

In my work with clients, I often find that companies start with Big Data technology rather than first identifying their business needs. As a result they spend months or even years building technology for technology's sake rather than solving a specific business problem.

One way to address this common pitfall is to get business users and technologists together very early in the process. Quite often these individuals are in silos; by breaking down those silos and getting them together sooner than later, you are far more likely to solve a well-defined business need.

For entrepreneurs—those working independently and those working in larger organizations—Big Data presents some amazing new opportunities. You can focus on a specific vertical such as healthcare, education, transportation, or finance, or you can build a general-purpose analytics application that makes Big Data more accessible. Uber is one example of a company that is using Big Data and mobile to disrupt the transportation and logistics markets. Meanwhile, Platfora is a new cloud-based visualization application. With ever-increasing interest in Big Data, the timing could not be better for building your own compelling application.

■ **Tip** Now is the time to build your own Big Data Application. For one thing, you'll learn a lot. For another, you can bet that your top competitors are hard at work building applications of their own.

Identifying Data Sources

Depending on the type of application you're building, data can come in many forms. For example, my company, Content Analytics, develops a product that analyzes entire web sites. Data comes from crawling tens of millions of web pages, much like Google does. The company's product then extracts relevant data such as pricing, brand segmentation, and content richness information from each of the web pages.

While identifying your data sources, it's also important to determine whether those data sources are unstructured, structured, or semi-structured. As an example, some web pages contain text and images almost exclusively. This is considered unstructured data.

Many e-commerce web sites contain pricing, product, and brand information. However, such data, as presented on the web pages, is not available in a well-known structured form from a database. As a result, the data has some structure but is not truly well-structured, making it relatively difficult to parse. This is considered semi-structured data.

Finally, data such as customer contact information stored in a database is a good example of structured data. The type of data you'll be working with will have a big impact on how you choose to store that data and how much you have to process the data before you can analyze it.

Data can also come from internal sources. One telecommunications client we worked with came up with the idea of developing a new Big Data field-management application. The application would bring together data about customer support requests with field technician availability and even customer account value to determine where technicians should go next. It could also incorporate mobile/location information to remind technicians to bring the proper equipment with them when they arrive at each service call. Major telecommunications companies have upwards of 75,000 or more field technicians. As a result, using data to deploy those technicians efficiently could potentially save a company hundreds of millions of dollars on an annual basis.

In another example, data can come from a company's web site. Site performance data and visitor information are both available, but need to be stored, analyzed, and presented in a compelling way. It's not enough just to store the log files a web site generates. Companies like AppDynamics and New Relic have created entire new businesses around application performance monitoring, while products like Adobe Marketing Cloud and Google Analytics convert massive amounts of web site traffic data into insights about who's visiting your site, when, and for how long.

Finally, don't forget public data sources. From economic data to population data, employment numbers and regulatory filings, an immense amount of data is available. The challenge is that it is often difficult to access.

Real estate information web site Zillow built an entire business from difficult-to-access real estate sale price information. Such information was accessible in public records, but those records were often stored in paper format in local town clerks offices. If such records were stored electronically, they often were not accessible online. Zillow turned this hard-to-access data into a central repository in standardized form accessible via an easy-to-use web interface. Consumers can enter an address and get an estimated price based on sales data for the surrounding area. Public data can be an invaluable resource. Sometimes you just have to do the work necessary to obtain it, organize it, and present it.

Once you have identified both the business need and your sources and types of data, you're ready to start designing your Big Data Application. A typical Big Data Application will have three tiers: infrastructure to store the data, an analytics tier to extract meaning from the data, and a presentation layer.

Choosing the Right Infrastructure

For most startups today, the decision to store and process their Big Data in the cloud is an easy one. The cloud offers a low up-front time and cost investment and on-demand scalability. In terms of how to store the data, there are typically three options. You can use the traditional SQL (Structured Query Language) database infrastructure. This includes products like MySQL, Oracle, PostgreSQL, and DB2. To a certain scale, MySQL works remarkably well and offers an unbeatable price—it's free. It also has the benefit of running on commodity hardware. There are also cloud-based versions of these databases available. Amazon, for example, offers its RDS service, a per-built, cloud-based data store that has features like redundancy built-in.

However, storing your data in a traditional database carries with it a range of challenges, not least of which is that the data has to be well structured before it can be stored. For many applications, the database tends to become the bottleneck fairly quickly. Too many tables that keep growing with too many rows, a tendency to scale up—that is, to add more processing, storage, and memory capacity on a single machine rather than splitting data across multiple machines—and poorly written SQL queries can cause even mid-size databases to grind to a halt. When it comes to Big Data, the application itself is often the least complex part—the data store and the approaches used to access it are often the most critical aspect.

At the other end of the spectrum are NoSQL databases. NoSQL databases overcome many of the limitations of traditional databases—the complexity of splitting data across multiple servers and the need to have structured defined up-front. However, they suffer from some severe limitations. Specifically, they have limited support for traditional SQL query-based approaches to storing and retrieving data. Recently, NoSQL databases have added more SQL-like capabilities to address these limitations, and they certainly excel at storing large quantities of unstructured data, such as documents and other forms of text. There are a number of products available in this area, but the most well-known (and best funded one) is MongoDB.

Finally, there is what has become the mainstream Big Data solution, Hadoop. The Apache Hadoop distribution is available for free and the design of the technology makes it easy to distribute data across many machines for analysis. This means that instead of having to rely on one database located on one or a few machines, data can be spread across dozens if not hundreds or thousands of machines in a cluster for analysis. For those needing commercial support, more advanced management capabilities, and faster or near-real time analytics, commercial solutions are available from Cloudera, HortonWorks, MapR, and other vendors.

If your application is cloud-based, you can either configure your servers for Hadoop and MapReduce or, depending on your vendor, use a pre-built cloud-based solution. Amazon MapReduce (EMR) is one such service that is easy to use. Amazon also offers the AWS Data Pipeline service, which helps manage data flows to and from disparate locations. For example, the Data Pipeline could automatically move log files from a cluster of server instances to the Amazon S3 storage service and then launch an EMR job to process the data.

The downside of using such pre-built, platform-specific solutions is that it makes it somewhat harder to move to another cloud platform or a custom configuration should you want to do so later. But in many cases, the benefits of using such solutions, such as the speed with which you can get started and the reduced maintenance overhead, often outweigh the potential downsides.

Note Keep in mind that when you go with a specific platform, like Amazon's version of MapReduce called EMR, you may have trouble migrating the data and all the work you put into creating a solution to another platform. Often, the benefits outweigh the risks, but give it some thought before you commit a lot of resources.

Presenting Insights to Customers

Once you have an approach to collecting, storing, and analyzing your data, you'll need a way to show the results to your users, customers, or clients and enable them to see the results and interact with the data. In many cases, their initial questions will beget more questions, so they will need a way to iterate on their analysis.

In terms of your application, you could design everything from scratch, but it makes a lot more sense to leverage the large number of pre-built modules, tools, and services available. Suppose your application needs to support the display of data on a map and the ability for users to interact with that data. You could build an entire mapping solution from scratch, but it is much more effective to make use of pre-built solutions.

If you're building a web-based data analytics service, CartoDB, BatchGeo, and Google Fusion are cloud-based data presentation and mapping solutions that make it easy to integrate geographic visualization into your application. Instead of writing a lot of custom visualization support, you simply upload your data in a supported format to one of those services and integrate a little bit of code into your application to enable the selected service to display the resulting visualization.

Such cloud-based mapping visualization solutions are also particularly useful for mobile applications. If you're building a fleet routing application, for example, using pre-built modules means you can focus on building the best possible algorithms for routing your fleet and the easiest-to-use application interface for your drivers, rather than having to worry about how to display detailed map data.

The downside to using such services is that your data is accessible to those services. While they may be contractually obligated not to use your data other than to show visualizations on your site, there is some added security risk when it comes to confidential data and it's important to weigh the tradeoffs between ease of use and security when deciding to use any third-party cloud-based visualization solution.

If your use case demands it, you can instead integrate open source or proprietary geographic visualization software directly into your application. CartoDB, for example, is available as an open source solution as well.

Pre-built modules are available to support a number of other visualization capabilities as well. HighCharts is one solution that makes it easy to add great-looking charts to your application.

Depending on the nature of your application, more or less custom presentation development work will be required. My company Content Analytics uses a combination of custom visualization capabilities and pre-built solutions like

HighCharts to provide complete visualization support in its product. This approach means the application's users get the best of both worlds: a specialized user interface for viewing data about online marketing effectiveness combined with elegant visualizations that display and chart the data in compelling and easy-to-understand ways.

Another approach to visualization is to focus on data analysis in your application and leave the presentation to products focused on visualization, such as Tableau, QlikView, and Spotfire. These products were designed for visualization and have extensive capabilities. If the competitive advantage for your application comes from your data and/or the analytics algorithms and approaches you develop, it may make most sense to focus on those, make it easy to export data from your application, and use one of these applications for data presentation. However, as a broader set of users, especially those without traditional data science and analytics backgrounds, integrate Big Data into their daily work, they will expect visualization capabilities to be built directly into Big Data Applications.

Gathering the Engineering Team

Building a compelling Big Data Application requires a wide range of skills. On your team, you'll need:

- Back-end software developers who can build a scalable application that supports very large quantities of data.

- Front-end engineers who can build compelling user interfaces with high attention to detail.

- Data analysts who evaluate the data and develop the algorithms used to correlate, summarize, and gain insights from the data. They may also suggest the most appropriate visualization approaches based on their experiences working with and presenting complex data sets.

- User interface designers who design your application's interface. You may need an overall designer who is responsible for your application's look and feel, combined with individual designers familiar with the interface norms for specific desktop, tablet, and mobile devices. For example, if you are building an iPhone interface for your application, you will need a designer familiar not only with mobile application design but also with the iPhone-specific design styles for icons, windows, buttons, and other common application elements.

- DevOps (Development-Operations) engineers who can bridge the worlds of software coding and IT operations. They configure and maintain your application and its infrastructure, whether it is cloud-based or on-premise. DevOps engineers usually have a mix of systems administration and development expertise, enabling them to make direct changes to an application's code if necessary, while building the scripts and other system configurations necessary to operate and run your application.

- A product or program manager whose mission it is to focus on the customer's needs. With Big Data Applications it is all too easy to get lost in technology and data rather than staying focused on the insights that customers and users actually need.

Combine the right team with the right data sources and technologies, and you'll be able to build an incredibly compelling Big Data Application. From fleet management to marketing analytics, from education to healthcare, the possibilities for new Big Data Applications are almost unlimited.

Doing a Big Data Project

Start with the End in Mind

When it comes to Big Data, it's best, as the old saying goes, to start with the end in mind. With Big Data, it's easy to get lost in the data or the implementation. The outcome you want to achieve and the business questions you want to answer are what matter most. So the best place to start in Big Data is not with the data itself, but rather with your business objectives and questions.

When setting up your Big Data project, you'll want to:

- Define the outcome
- Identify the questions to answer
- Create policies around what data you use and how you use it
- Measure the value
- Identify the resources, including the people, data, hardware, and software, that you need to answer those questions
- Visualize the results

Define the Outcome

It's all too easy to get lost in technology and data in the world of Big Data. That's why it's critical to define your desired business outcome at the start of any Big Data project. The most well-defined outcomes are highly measurable in terms of business value. Here are a few examples:

- Increase the Net Promoter Score (NPS), which is a frequently used measure of customer satisfaction, by three points within six months and then maintain that NPS score going forward.

- Increase sales of Product X by 10% in one week through higher advertising conversion rates.

- Reduce customer churn by 2% in two months by improving customer satisfaction.

- Increase high school graduation rates by 10% in three years.

- Decrease emergency responder response times by 30 seconds per response in one year.

- Achieve sub-second web site response time for initial application login in three months, and then continue to deliver sub-second response time at least 98% of the time.

- Gain sufficient funding for a new project by analyzing key sources of data and demonstrating a clear market need.

- Reach 100,000 active users for a new application within three months of launch.

- Increase user engagement on a web site by two minutes per day over the course of a six-month period, and then maintain that increased engagement rate for at least 75% of all active users.

Notice that many of these objectives define both an initial target and then an on-going target. This is important because a one-time improvement may be a lot easier than a sustainable improvement. Unless both objectives are identified, those responsible for the implementation might choose a short-term fix (such as adding memory to a single server) over a long-term fix (such as redesigning an application so that it can run on multiple servers).

If you're not sure of the exact business metrics you want to achieve, but you know that you have a need, you might start with more general objectives, like the following:

- Create a real-time visualization of security attacks on corporate assets (both online and physical).
- Enable technicians to see attacks as they happen.
- Improve the speed with which technicians respond to security attacks.

A refined version of this objective would put specific numbers on the percentage of attacks successfully displayed in visual format, how long such data takes to appear in visual form, and how quickly technicians are able to respond. Such numbers would come after an analysis of current security attack rates and response times.

These are just a few of the many potential business objectives you could identify for your Big Data project. The more specific you can make your business objective, the easier it is to stay focused and achieve that objective in a timely manner. When you can put a value on the objective you want to achieve, it becomes that much easier to obtain necessary resources such as funding, people, and access to data sources.

Identify the Questions to Answer

Much like defining your outcome, you should also prepare a list of questions you want to answer. If you don't have baseline metrics readily available, your first set of questions might involve some basic data gathering. For example:

- What is the average web page load time for the checkout page of our e-commerce web site?
- What is the best time? What is the worst time?
- What are the industry norms or best practices for web site response times?
- Are we above or below those numbers?
- If we achieve sub-second web site response time for the checkout page of our site, what increase in sales should we expect?

Once you have a baseline set of data, you then iterate on your question set, as follows:

- What products can we add to the mix of products on our home page in order to increase sales?

- Do certain product descriptions or images result in higher conversion rates?

- What kinds of recommendations to the customer are most effective in increasing basket size (the size of their ultimate purchase)?

These kinds of questions can only be answered through an iterative process of experimentation. By testing different approaches, featured products, product images, and other aspects of a site, you can learn what works best in producing the most relevant product offers for your potential customers—and the most sales for your business.

This approach applies to other Big Data areas as well. You might run comparison experiments to evaluate the impact on customer satisfaction of human responses to customer services inquiries versus automated email responses. There are lots of different ways to put Big Data to work, from building entirely new applications to visualizing what's going on so you can make better decisions. The key is to take an explorative, discovery-oriented approach to your Big Data project. The faster you can iterate, the faster you can ask new questions based on what you learn from answering your previous set of questions.

Create Data Policies

It has been said that with great power comes great responsibility. Data provides enormous power and thus brings with it enormous responsibility.

When undertaking your Big Data project, it's critical to define policies for your data use. Consider the value exchange between data collected and value delivered. If you give your customers better product recommendations and deals based on your data analytics, your customers are likely to thank you. But if you market baby products to your potential customers before they even know they're pregnant, as one major retailer did, you will very likely find yourself on the wrong side of the data-value equation.

One good rule of thumb is to ask yourself two simple questions. First, would you want your own data used in the way you plan to use it? Second, is the way you're planning to use the data you collect good for your customers? At a minimum, asking such questions at the outset of a Big Data project will start an important discussion about data use and policy.

In addition to data policy, you'll want to consider the security of your data—both in motion and at rest. Data in motion is data that is moving from one location to another. When you enter your credit card number into a web page and your web browser sends that number to a server somewhere on the other side of the country, your important, personal financial data is set in motion.

While it's traveling from your computer to the destination (such as an e-commerce site where you're placing an order for that new set of headphones you've had your eye on), your credit card information is at risk. Hackers can potentially access it and use it for their own purposes. Now consider this risk multiplied millions of times—the movement of personal data from one place to another or the movement of medical records from one doctor's office to another. If you're working with data on a large scale, especially if that data contains personal information, considering both the policies around the use of that data and the approaches to securing that data is critically important.

Once the data arrives at its destination, the data is "at rest." Such data is typically on-disk, stored in a server's file system or a database, or in the memory of the computer that's processing it. Which data needs to be encrypted and what approaches should be used for doing so are also important questions to answer. On the surface it seems like you might want to encrypt all data. In practice, encrypting and decrypting data is both time consuming and consumptive of valuable computer resources. As a result, the vast majority of data is not encrypted. These kinds of considerations become more and more important the larger the scale of data you're working with and the bigger your company brand becomes. The company with the biggest brand is often the one most likely to suffer the consequences if a data breach occurs—just consider the recent exposure of tens of millions of credit card records from retailer Target. While there were other partners and vendors involved, Target itself suffered the most visible consequences.

Measure the Value

Business value can be measured in multiple ways. There is financial impact, of course, such as the amount of money saved by making a business process more efficient or the additional revenue generated from a successful marketing campaign. But you can measure the value of Big Data in other ways, such as its ability to promote innovation or support risk-taking in the development of new projects.

Assuming you've identified the key objectives for your project at the outset, it is then simply a matter of measuring those key metrics as your project proceeds and tying those metrics back to revenue or cost savings. For example, you can:

- Measure web site response time and see if there is a corresponding increase in sales, and if so, by how much.

- Compare one version of your home page with another version of it and determine which produces the best results. You can then look at the percentage or dollar value increase in sales and measure the impact of your changes.

- Evaluate the customer satisfaction levels of your customers, reduce the amount of time it takes to respond to service requests, and then re-evaluate customer satisfaction levels. If you know how much a certain increase in customer satisfaction reduces customer churn, and what an average customer is worth to you, you can measure the business value of reducing customer support response times.

The keys to measuring the value are incorporating on-going measurement tools and having financial or other metrics you can tie those measurements back to. On-going measurement is critical because any measurement you take only once might not actually be representative of what's really going on. Bad decisions could result. By taking regular measurements, you can see how the results change over time—which makes it easier to recognize any potentially invalid data as well as to make sure you're headed in the right direction.

Identify Resources

Beyond human capital, you need three kinds of resources for your Big Data project:

- Sensors for collecting the data

- The data itself

- The necessary tools to analyze the data and gain actionable insights

Many kinds of hardware and software generate and collect data. Global Positioning System (GPS) watches record location and elevation data, and by connecting to other devices, they also collect heart rate and cadence information. Airplane engines generate data about fuel consumption and engine performance, which is collected using software. Meanwhile, mobile phones generate many data points every second, from call signal strength to power consumption.

Mobile phones don't just generate data about themselves, they can also gather data, such as a user's location. With the right software and hardware, they can also monitor heart rate, sleep patterns, and a wide variety of other health information, such as EKG data and blood sugar levels. This can help people improve their day-to-day health and also provide lower-cost measurement alternatives for diabetics and other patients.

In many cases, you don't need to search far to find the data you need. Lots of data already exists, such as call center, billing, customer service, and product usage data. Marketing companies have data about marketing spend, sales, and margins. Human resource organizations have data in the form of resumes, retention length, employee ratings, and compensation.

The data you want to work with often isn't in the form you need it. For example, call centers or sales department may record calls for quality assurance or training purposes, but making sense of that data is hard when it remains in audio form. By converting these recordings into text form, organizations can then analyze the text and discover trends and insights to which they would not otherwise have had access.

Note Data comes in many forms. It can be data from your server logs or email inquiries, or it might be data picked up by sensors—think of a heart monitor or a new car's ability to monitor and store all manner of electrical, mechanical, or driver data. The common need, however, is the ability to capture data from such sources and use it to spot trends or uncover insights about customers that lead to new products or better service.

Similarly, web applications may have lots of data, but that data can be difficult to unlock without the right tools. For example, if your web site is performing slowly, you want to find the root cause of that performance slowdown. The right software can log a stream of data points about web site performance as well as about sources of potential underlying performance bottlenecks such as databases and servers.

A frequent stumbling block for organizations brainstorming new Big Data projects is the question of how to get the data. Quite often, projects that might otherwise lead to meaningful insights are stopped before they start. For example, a mobile application developer might want to build an application to reduce food waste by alerting people to grocery purchases that they consume—and those that are thrown away unused.

The financial savings from such an application have the potential to be enormous while the societal impact could be profoundly positive, with less food being wasted and more going to those in need. Yet how would the application developer get access to the necessary consumer grocery purchase data and to data about which products the consumer uses versus throws away?

It would be all too easy to get stopped by such questions. But the value to be gained may more than justify the investment required to access and analyze such data and to build a compelling mobile application that consumers would love to adopt. By starting with a focus on the business value rather than on the data itself, it becomes far easier to achieve your Big Data project goals.

Regardless of what data you choose to work with, you'll need to store that data somewhere so that you can analyze it, either locally or in the cloud. If your company has existing Information Technology (IT) resources, you may be able to get the servers, storage, and processing capabilities you need internally. However, IT departments are often busy with other tasks. In such cases, or if you're not sure up-front how much data you'll need to store or how much analytics processing you'll be doing, storing and analyzing your data in the cloud makes more sense.

Your choice of technology also depends on whether you're working with structured or unstructured data. Traditional databases may be more than adequate for storing contact information, billing records, and other well-structured information. Although unstructured data, like web pages and text files, can be stored directly on-disk, when it comes to working with such data at scale, more sophisticated technologies are required. This is where technologies like Hadoop and MapReduce come into play.

No matter how much data you crunch, at the end of the day, you'll need to make your analysis accessible and actionable. This is where visualization products come in. Offerings from companies like Tableau Software and QlikTech make it easy to take vast amounts of data and create compelling interactive visualizations.

Visualize the Results

Putting your analysis into visual form is one of the most effective ways to make your Big Data project actionable. Often when people think about Big Data, they think about indecipherable spreadsheets or the streams of 1s and 0s flowing down the screen in *The Matrix*. Visualization turns all those 1s and 0s into a clear picture that anyone can understand. It makes the unapproachable approachable.

It may seem like a lot of different capabilities are required to work with Big Data. Yet working with Big Data really comes down to just three things: sensors, data, and software.

To make things even simpler, new Big Data Applications (BDAs) are emerging that take all of the pieces—collecting the data, storing the data and presenting the data—and put them together into easy-to-use applications. In many cases, these BDAs run in the cloud and are delivered over the Internet, meaning you don't need to buy any hardware or install any software to get your Big Data project started. The Big Data Landscape (http://www.bigdatalandscape.com/) highlights some of the companies whose offerings you may want to look at depending on your area. Examples include cloud-based Big Data offerings from companies like AppDynamics, CartoDB, and New Relic, to name just a few.

In many cases, even if you have what may seem like a custom Big Data problem on your hands, it is well worth looking for a Big Data Application before deciding to build your own. There are BDAs that do everything from web site performance analysis to fleet management, many of which can be customized to meet your specific requirements.

Case Study: Churn Reduction

Over the next seven years, the market for Big Data technology in the telecommunications sector is expected to grow some 400%, according to industry analyst Ari Banerjee at Heavy Reading. That's[1] not surprising given the tremendous amount of data that telecommunications companies work with. From mobile phones to Internet connections, from call centers to data centers, telecommunications companies work with a lot of data. Network data, service data, and subscriber data are just a few of the many data sources that telecommunications companies work with.

The Big Data use cases in telecommunications are numerous. They include:

- Churn management
- Customer experience management
- Handset analysis
- Roaming management
- Revenue management
- Interconnect billing verification
- Content settlement assurance
- Fleet routing management
- E-commerce

---

[1]http://www.lightreading.com/spit-(service-provider-it)/analytics-big-data/telco-big-data-market-to-thrive/d/d-id/707393

Each of these areas presents a tremendous Big Data opportunity for today's telecommunications companies, an opportunity not just to analyze data for data's sake but to serve their customers better.

For example, analysis of handset performance can lead to longer battery life and higher reliability. It can also enable call center operators to troubleshoot customer issues. With the right data, rather than relying on a customer to describe a perceived problem, a call center operator can easily determine whether a problem is in a customer's handset or tablet, in one of the applications installed on the customer's phone, or in the network itself.

Big Data Workflow

To analyze the data and determine which actions to take, I will use the following five-step workflow (see Figure 7-1), which includes creating a business hypothesis, setting up the necessary systems, transforming the data, analyzing and visualizing the data, and acting on the results.

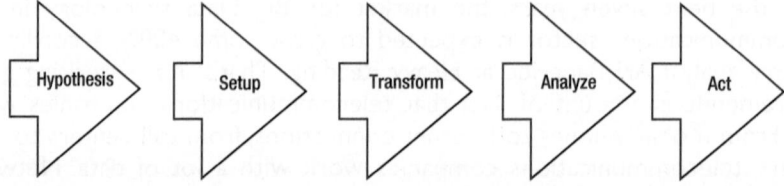

Figure 7-1. The Big Data workflow

The example we'll cover in detail here is churn management. Churn is a challenge for nearly every company. No matter the size of your business, losing a customer—having a customer "churn out"—is an expensive event. In most cases, it's more cost effective to keep a customer than add a new one.

To reduce churn, it's important to predict which customers are going to churn and execute proactive retention campaigns to keep them. In this example, we'll use two technologies for the analysis—a data-cleansing tool and Hadoop MapReduce for storing and analyzing the data.

Step 1: Create a Hypothesis

One way to predict which users are going to churn is through behavioral analysis. With this approach, we try to find correlations between demographics, usage locations, call experience, and consumer behavior. By modeling which of these data points consistently occur together, we can predict customers who are likely to stay and those who are likely to churn out.

Our hypothesis is that subscribers who experience high rates of dropped calls are more likely to leave our service.

If our hypothesis proves to be true, we can then find all customers who experience high rates of dropped calls and take action to improve their experience.

For example, we could determine where large numbers of customers with dropped calls are geographically located. With that information, we could then investigate the root cause of the dropped calls. A high rate of dropped calls could be due to a faulty piece of network equipment that needs to be replaced. Or, our network might be overcapacity in certain areas and we could then add more towers or network bandwidth in those areas to reduce dropped call rates.

For customers who are experiencing a high volume of dropped calls but are not grouped with a large volume of other customers, we could determine if a different root cause is the issue, such as a problem with a particular customer's mobile phone or installed applications that are causing that customer's device to drop calls.

Step 2: Set Up the Systems

The first step is to gather the relevant data sources. In this example, we have two important data sources we need to work with: Call Detail Records (CDRs) and subscriber details. CDRs are typically stored in raw log files that are not cleanly structured for analysis.

Here is an example of a CDR, showing a user calling into a voicemail system[2]:

```
"""Console""" <2565551212>","2565551212","*98","UserServices",
"Console/dsp","",
```

```
"VoiceMailMain","@shifteight.org","2010-08-16 01:08:44","2010-08-16
01:08:44","2010-08-16 01:08:53","9","9","ANSWERED","DOCUMENTATION","",
"1281935324.0","",0
```

The challenge for the data analyst is that call records are raw log files that are not cleanly structured for visualization. When it comes to Big Data, this is a common problem. According to one source, 80% of the work in any Big Data project is cleaning the data.[3]

---

[2]http://www.asteriskdocs.org/en/3rd_Edition/asterisk-book-html-chunk/
asterisk-SysAdmin-SECT-1.html
[3]http://radar.oreilly.com/2012/07/data-jujitsu.html

While we could easily analyze a few call records on a single computer, we need data from millions of call records to run our analysis. To do that, we'll first set up a Hadoop cluster.

To set up a Hadoop cluster, we'll need a set of servers—either existing or new—for the cluster and the necessary operating system and Hadoop software to run on those servers. We have two options for our Hadoop cluster: we can perform the setup and configuration on servers in an in-house, on-premise data center or we can use Hadoop resources available in a public cloud such as Amazon Elastic MapReduce (EMR).

In this case, because the data we want to work with is already in our on-premise data center, we'll set up our own Hadoop cluster in-house. (For Big Data Applications where data is already stored in a public cloud, such a clickstream analysis for a web site, it makes more sense to perform the data processing jobs in the cloud.)

Setting up a Hadoop cluster simply means installing the necessary operating system software on the servers, typically some form of Linux. We'll install the necessary Hadoop-related software and its Java dependencies. And we'll use a standardized deployment tool (such as Puppet) to automate the installation and setup of these software pieces so that it's easy to spin up additional servers as we need them.

Once we have this set up, we'll load the call records into Hadoop (assuming they were not stored there originally) and we'll be ready to run our data transformation.

Step 3: Transform the Data

Before we can analyze our data, we have to define and run a set of data transformations to get the data into a form we can analyze.

Transformations take the semi-structured data in the call records and turn them into a form that we can examine and make decisions about. Historically, data analysts have created such transformations by writing scripts in a painstaking, manual process. However, new tools are available that enable the analyst to view raw data in an easy-to-use interface and give the tools examples of the kinds of transformations that need to be done. In this case, we'll use a tool that uses the examples that we provide, along with machine learning, to gener-ate the necessary transformation scripts automatically. In this particular case, once users give the tool some initial examples of how to convert one data format into another, the tool can recognize similar types of data that need to be converted automatically. Over time, as the tool builds up more rules and more patterns it can recognize, it becomes easier for the software to make the necessary transformations without human intervention.

For example, one transformation involves removing data that is not relevant to our analysis. In telco systems, both text messages and calls typically result in call records. So in the case of call records, that means taking out the text messages so that we can analyze just the relevant data, that is, the records related to actual calls.

A second transformation involves combining the subscriber data with call record data so that we can view correlations between dropped calls and various subscriber attributes such as geographic location.

The Hadoop Distributed File System (HDFS) stores all of the relevant data while MapReduce executes the necessary transformations on the data. What makes MapReduce so powerful is that it takes huge volumes of data, in this case the call records, and splits that data into smaller chunks that can be processed in parallel. As a result, by adding more servers to the cluster, we can increase processing speed and reduce the amount of time it takes to transform the data into a form we can analyze.

Step 4: Analyze the Data

Once the data transformation step is done, we can analyze the data to see if our hypothesis is correct. In this example, we created a dashboard view of the data in a popular visualization tool called Tableau (see Figure 7-2). (Many other visualization tools are available, including software from companies like QlikTech and MicroStrategy.)

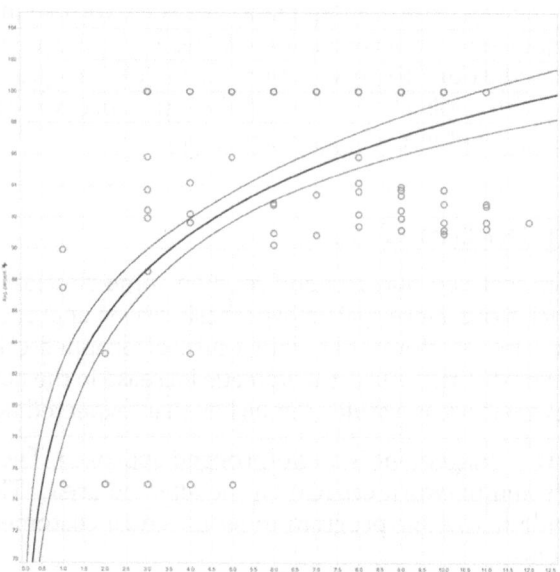

Figure 7-2. Visualizing the correlation between dropped calls and subscriber churn. (Courtesy of Tableau Software and Trifacta; used with permission)

Visualization is one of the most powerful ways to analyze data. Not only do our eyes provide the highest bandwidth data transfer to our brains of all of our senses, but data shown visually is simply easier to interpret than rows upon rows of spreadsheets or databases. Often, massive amounts of data can be converted into simple charts that make it easy to draw powerful conclusions.

■ **Note**   Our own eyes provide the highest-bandwidth data transfer to our brains compared to all our other senses. That's why data visualization is such an important part of a Big Data project. It provides actionable information the fastest.

Figure 7-2 shows that customers with higher successful call completion rates remain subscribers longer than those with more dropped calls. As a result, we can conclude that there is a correlation between higher dropped call rates and higher subscriber churn.

It's important to keep in mind that there could be other factors at work that are causing our customers to churn out that we haven't yet evaluated. These factors could include unattractive pricing, poor customer service, or strong marketing campaigns from our competitors. We'll want to explore these factors in future analyses. Our initial analysis also brings up additional questions. We'll want to dig into the root cause of the dropped calls—is it an issue with the devices, the network, or something else altogether?

This is the compelling nature of Big Data. We start with an initial hypothesis and a set of questions to answer, analyze the data, and then develop new hypotheses and questions to answer. Because we can analyze large quantities of data quickly with tools like Hadoop and MapReduce, we can take a much more iterative and exploratory approach to our data.

Step 5: Act on the Data

Now that we've analyzed our data and reached some conclusions, it's time to act on our conclusions. Here again a hypothesis-driven approach is the way to go. In this case, after exploring the root cause of dropped calls in particular areas, we discover that the cause is a dramatic increase in the number of callers in certain geographic areas, resulting in higher than expected network usage.

Our proposal to management is a two-pronged approach. The first step is to add call towers and network capacity in the affected areas. The second step is to implement a marketing program to let potential customers know about the improvements.

Part of our marketing program will be an email campaign targeting previous subscribers who churned out with an offer to come back and news about the increased network capacity in their area. The other part of our marketing program is a billboard-based campaign to tell potential new subscribers about the increased network capacity in their area.

In both cases, we can use data to implement the most effective campaigns possible. For example, in the case of the email campaign, we can try multiple different email campaigns, such as different wording and offers, to determine which campaign is most effective. Of course, we'll use all the techniques we've learned so far about Big Data when we implement those campaigns. We'll take a look at effective Big Data approaches to marketing analytics in the very next section.

Case Study: Marketing Analytics

So that we can explore other markets, we'll continue our Big Data project by switching to the area of e-commerce. Regardless of the market, however, many of the same principles apply.

In this example, an e-commerce company wants to predict the performance of a new product launch they are about to do and ensure the highest conversion rates and sales possible once the launch is underway. Of course, the scenario could apply equally well to the launch of a new mobile device for a telecommunications company or a new web site service for a technology company. In this case, an online retailer is launching a new women's clothing line for the summer.

The campaign is scheduled to launch in just one week and the retailer wants to test the marketing campaign and see which products will be best sellers. By doing so, they can better prepare and optimize inventory and distribution.

At the same time, the retailer wants to evaluate the effectiveness of different email campaigns. To do this, they send different emails to 10% of their customers over the week to see which emails produce the highest conversion rates; their challenge is monitoring the campaign results in real time.

Before starting the study, the retailer defines a set of key questions to answer. These include:

- Which products will be the top sellers?
- Which email campaigns resulted in the most sales?
- Can we optimize the web site to improve conversion rates?

The marketing team faces a number of challenges. The team has only one week to test the launch in its chosen test market. They want to iterate quickly, but existing Business Intelligence (BI) systems only provide daily reports. This speed is not fast enough for the marketing team to make changes and test the results given the time they have available before the full launch. The team wants to be able to iterate several times a day to make the most of the time available.

With Big Data, these challenges are easily overcome. Unlike the on-premise Hadoop-based solution that we looked at in the case of churn reduction, in this case we're going to look at a leading edge cloud-based Big Data approach that requires no hardware or software setup. That means that team doesn't need to purchase or configure local servers, storage systems, or software. All the necessary capabilities are available over the Internet via the cloud.

What's more, with this approach, there's no batch-driven processing as in the Hadoop-based approach. With this cloud-based Big Data approach, the retailer can get real-time reports, immediate insight into the correlation between different email campaigns and consumer purchase behavior, and the ability to test the effectiveness of web site changes and a variety of different banner ads.

Big Data Meets the Cloud

How does this work? A variety of marketing analytics products are available. In this case study, we explore the use of New Relic Insights, a relatively new offering for analyzing streams of online marketing data and getting real-time insights.

By way of context, products such as New Relic, AppDynamics, and others have traditionally focused on the Application Performance Monitoring (APM) market. That is the on-going measurement and analysis of bottlenecks in web site and enterprise application performance.

These products are the cloud-based alternatives to products historically offered by enterprise vendors such as Computer Associates, BMC, and others. More recently, these cloud-based solutions have expanded their platforms to offer a broader set of Big Data capabilities, such as marketing analytics. They leverage the same core capabilities that have allowed them to stream in, store, and analyze immense quantities of web site performance data to provide insight into other business areas.

This cloud-based approach to computing was pioneered by the "no software" company Salesforce.com. Salesforce.com took a cloud-based approach to Customer Relationship Management (CRM), which had traditionally only been available as part of expensive, custom on-premise software implementations from vendors like Siebel Systems (now Oracle).

There are some tradeoffs with this approach. In particular, your data is stored on a vendor's computer systems rather than on your own. But in reality, vendors like Salesforce.com have architected their systems to manage the data of thousands and thousands of customers reliably and securely. And because their services are available in the cloud, there's no expensive up-front investment or time-consuming setup. The vendor handles all the software configuration and maintenance so that you can focus on getting the results you need.

With the cloud-based approach to Big Data there are just three steps in the workflow (see Figure 7-3).

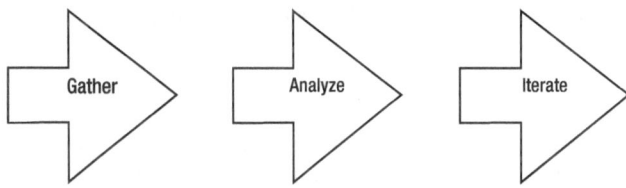

Figure 7-3. The Big Data cloud workflow

Step 1: Gather the Data

To accomplish its goals, the marketing team in this case study simply needs to add a few lines of JavaScript to the company's existing web pages, and the data to analyze campaign performance is uploaded to the third-party, cloud-based application. From there, members of the marketing and merchandizing teams can all log in and see the results.

The cloud vendor deals with the complexities of infrastructure, data capture, and report generation, while the retailer can stay focused on delivering the best results possible from its new product launch.

Step 2: Analyze the Results

In this example, the retailer is trying three different email campaigns to see which one is most effective. Each email campaign has different messaging and offers. The retailer can view product purchases by email campaign in real time using easy-to-create web-based dashboards, as shown in Figure 7-4.

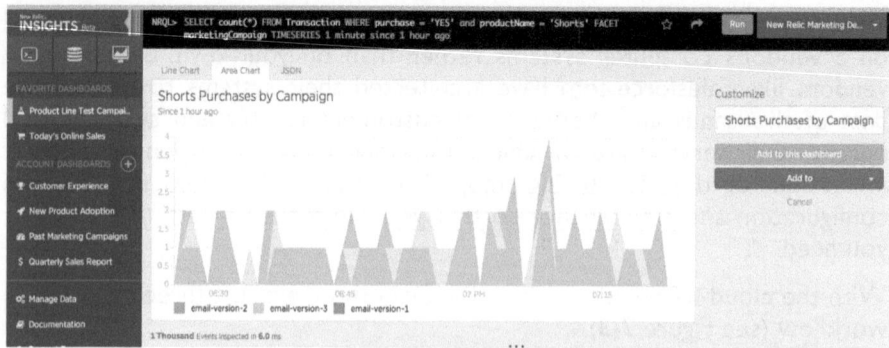

Figure 7-4. Purchase behavior based on different email campaigns. (Courtesy of New Relic; used with permission)

The retailer can also create more advanced, custom queries. Traditionally, only database administrators and data analysts have had the knowledge and skills to create queries directly on data and turn the results of those queries into meaningful reports.

New tools provide easy-to-access query capabilities that allow marketers to query data directly and turn those results into actionable charts easily. In Figure 7-5, marketers at the e-commerce site have entered SQL-like queries directly into the web interface to evaluate the performance of different banner ads.

Figure 7-5. Visualizing the impact of different banner ads on product purchase behavior. (Courtesy of New Relic; used with permission)

In this instance, the retailer is comparing the purchase behavior of consumers in response to two different banner ad approaches.

Step 3: Iterate

Once the test phase is complete, the retailer can continue to iterate and optimize the actual launch. For example, the retailer can test and optimize specific parts of its web site to improve engagement and conversion rates in real time using A/B testing.

With A/B testing, the different combinations of buttons, colors, fonts, and other web sites elements are shown to each visitor to the site. The optimization software automatically figures out which combination produces the highest conversion rates based on pre-defined goals, such as visitors clicking on a "register now" link or completing a purchase.

In this case, the retailer can use web-based Big Data tools such as Optimizely or Google Site Analyzer to try different versions of key pages on its web site to see which combination of content, products, offers, fonts, and colors produce the best conversion results.

But the changes don't stop there. The retailer can even build functionality directly into its web site that uses the output of the cloud-based analytics tool to change application behavior. For example, the site could be designed to promote particular products on the home page that are not selling as well as they should be. The site could also kick off email campaigns to promote certain products automatically or provide special offers at checkout.

A cloud-based approach to your Big Data project can make a lot of sense, especially for sales and marketing-related analytics. There's no expensive up-front investment required and no on-going maintenance. Continuous iteration is fast and easy.

Case Study: The Connected Car

We'll conclude this chapter on Big Data projects by looking at the way one company opened up access to a valuable yet previously closed data source—car data.

Cars may not seem like a huge source of data, but they are. In fact, our cars generate immense amounts of data not just about themselves but also about us as drivers. Ford wanted to understand the differences in driving behavior between drivers of gas and electric powered vehicles.

However, there were a number of challenges the company had to overcome to make analysis of car data possible. In particular, car data is typically difficult to access. A car may have a lot of data, but unlike a phone or a computer, there typically aren't any well-defined interfaces for software developers and business analysts to access that data.

Imagine that you wanted to build a mobile application to analyze your driving habits and help you drive most efficiently to conserve fuel/power while still getting you to your destination in time. How would you connect the app you built for your phone to your car?

Could you get the data in real time from the vehicle or would you need to download chunks of data after your drive? And how much data does a car generate? Is it so much that it would swamp the processing and storage capabilities of a mobile phone? If so, you might need to first send the data to a set of cloud servers for storage and analysis, and then download the results to your mobile application.

Yet if you could overcome these challenges, the benefits would be enormous. Car companies could improve engine efficiency by better understanding consumer driving habits, while drivers could change their driving habits to save gas while still getting where they want to go. What's more, if a manufacturer made it easy to access car data, software developers could create innovative applications for cars that no one has even thought of yet.

The single biggest challenge in the case of the connected car, however, was making it possible for developers to access the data. Ford developed the OpenXC interface, which is essentially an application programming interface (API) for car data. OpenXC makes it possible for developers to access car data in a standardized, well-defined way.

So instead of the application developers having to focus on how to get the data, they can focus on building innovative applications. Ford benefits because it gains insight into driver behavior as well as exciting new applications built for its cars. Consumers benefit because they can learn more about their own driving habits and take advantage of applications that can save them money or shorten their drive time.

Figure 7-6 shows one example of the kind of data that the OpenXC interface made available for a driving test, in this case a comparison of the driving habits of three different drivers. This is just one of the many kinds of data available via the OpenXC interface.

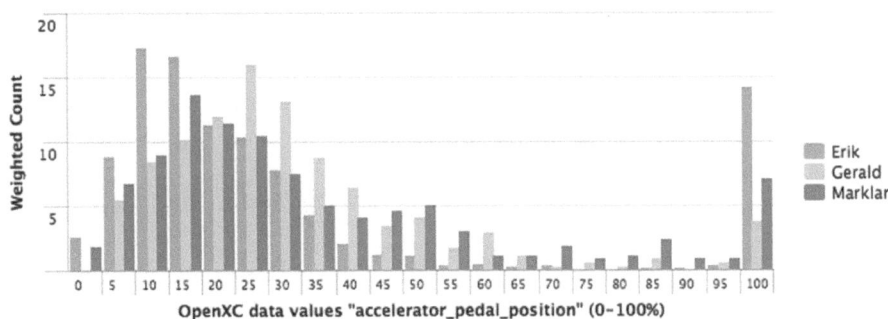

Figure 7-6. A comparison of the driving habits of multiple drivers. (Courtesy of Splunk; used with permission)

The key takeaway from the Connected Car example is that by opening up data access to their platforms, companies can create powerful ecosystems for innovation. In the case of the connected car, that means new applications that can help both consumers and manufacturers. Many other industries can also benefit from providing standardized interfaces to the data their systems produce.

Providing standardized access does not mean that anyone can access anyone else's data. Rather, it makes it much easier for developers to build innovative applications that can analyze data from all different kinds of systems, from medical devices to weather reporting stations.

In Short . . .

By combining the right tools, resources, and policies, you have the opportunity to gain incredible insights and unlock significant customer value using Big Data. Getting a Big Data project started need not be time-consuming or expensive. It can be a rapid, highly iterative undertaking that has the potential to deliver happier customers and unlock incredible value as a result.

The Next Billion-Dollar IPO: Big Data Entrepreneurship

Drive Data Like Billy Beane

As the General Manager of the Oakland Athletics, Billy Beane faced a problem. Beane and the Oakland A's had to field a competitive team without the big market budget of a team like the New York Yankees. In a story made famous by the movie *Moneyball* and the book of the same name by Michael Lewis, Beane turned to data and statistics.

This approach is known as sabermetrics, a term coined by baseball writer and statistician Bill James. Sabermetrics is a derivative of SABR, which stands for the Society for American Baseball Research. It is "the search for objective knowledge about baseball."[1]

---

[1] http://seanlahman.com/baseball-archive/sabermetrics/sabermetric-manifesto/

At the time, Beane's data-driven approach was widely criticized. It broke with years of the tradition of relying on the qualitative observations that scouts made about players. Instead it focused on using payroll to buy enough runs, not from a single player, but in aggregate, to buy wins.

Beane didn't use Big Data as it has traditionally been defined—as working with a volume of data bigger than the volume that traditional databases can handle. But he did apply data in a novel way to achieve a powerful impact.

Rather than making decisions based purely on qualitative information, Beane used a data-driven approach. More specifically, as Nate Silver points out in his best-seller *The Signal and the Noise: Why Most Predictions Fail but Some Don't*, Beane was disciplined about using the statistics that mattered, like on base percentage, rather than those that didn't.[2]

Silver developed the PECOTA system, short for Player Empirical Comparison and Optimization Test Algorithm, a sabermetric system widely used for forecasting the performance of major league baseball players. The PECOTA system draws on a database of some 20,000 major league batters since World War II, as well as 15,000 minor league batters.[3]

Today, those who want to take a data-driven approach in their own organizations often face the same challenge that Beane faced. Taking a data-driven approach to decision making isn't easy. But as Billy Beane showed, it does produce results.

Organizations know they need to become more data-driven, but to do so, it has to be a lot easier for them to *be* data-driven. Big Data Applications (BDAs) are one of the key advances that make such an approach possible. That's one of the key reasons BDAs are poised to create tech's next billion dollar IPOs.

Why Being Data-Driven Is Hard

Amazon, Google, IBM, and Oracle, to name a handful of the most valuable data-related companies on the planet, have shown the value of leveraging Big Data. Amazon serves billions of e-commerce transactions, Google handles billions of searches, and IBM and Oracle offer database software and applications designed for storing and working with huge amounts of data. Big Data means big dollars.

---

[2]http://bigdata.pervasive.com/Blog/Big-Data-Blog/EntryId/1123/It-s-Not-about-being-Data-Driven.aspx
[3]http://en.wikipedia.org/wiki/PECOTA

Yet baseball isn't the only game in town that struggles to make decisions based on data. Most organizations still find making data-driven decisions difficult. In some cases, organizations simply don't have the data. Until recently, for example, it was hard to get comprehensive data on marketing activities. It was one thing to figure out how much you were spending on marketing, but it was another to correlate that marketing investment with actual sales. Thus the age-old marketing adage, "I know that half of my advertising budget is wasted, but I'm not sure which half."

Now, however, people are capturing data critical to understanding customers, supply chain, and machine performance, from network servers to cars to airplanes, as well as many other critical business indicators. The big challenge is no longer capturing the data, it's making sense of it.

For decades, making sense of data has been the province of data analysts, statisticians, and PhDs. Not only did a business line manager have to wait for IT to get access to key data, she then had to wait for an analyst to pull it all together and make sense of it. The promise of BDAs is the ability not just to capture data but to act on it, without requiring a set of tools that only statisticians can use. By making data more accessible, BDAs will enable organizations, one line of business at a time, to become more data-driven.

Yet even when you have the data and the tools to act on it, doing so remains difficult. Having an opinion is easy. Having conviction is hard.

As Warren Buffet once famously said, "Be fearful when others are greedy and greedy when others are fearful."[4] Buffet is well-known for his data-driven approach to investing. Yet despite historical evidence that doing so is a bad idea, investors continue to invest on good news and sell on bad. Economists long assumed people made decisions based on logical rules, but in reality, they don't.[5]

Nobel prize winning psychologist Daniel Khaneman and his colleague Amos Tversky concluded that people often behave illogically. They give more weight to losses than to gains and vivid examples often have a bigger impact on their decision making than data, even if that data is more accurate.

To be data-driven, not only must you have the data and figure out which data is relevant, you then have to make decisions based on that data. To do that, you must have confidence and conviction: confidence in the data, and the conviction to make decisions based on it, even when popular opinion tells you otherwise. This is called the Big Data Action Loop (see Figure 8-1).

---

[4]http://www.nytimes.com/2008/10/17/opinion/17buffett.html?_r=0
[5]http://www.nytimes.com/2002/11/05/health/a-conversation-with-daniel-kahneman-on-profit-loss-and-the-mysteries-of-the-mind.html

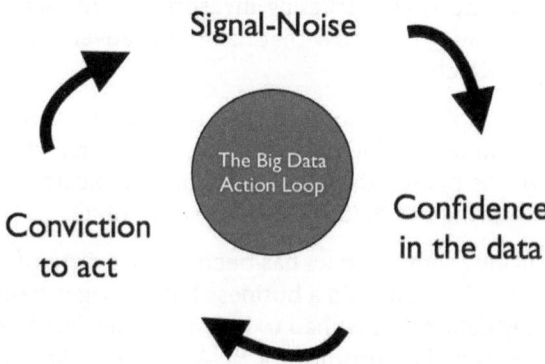

Figure 8-1. The Big Data Action Loop shows the key elements of Big Data decision-making

Hearing the Signal and Ignoring the Noise

The fact that being data-driven is so hard, both culturally and from an implementation perspective, is one of the key reasons that BDAs will play such an important role going forward. Historically, data was hard to get to and work with. Critically, data frequently wasn't in one place. Internal company data was spread across a variety of different databases, data stores, and file servers. External data was in market reports, on the web, and in other difficult-to-access sources.

The power, and the challenge, of Big Data is that it often brings all that data together in one place. That means the potential for greater insights from larger quantities of more relevant data—what engineers refer to as *signal*—but also for more *noise*, which is data that isn't relevant to producing insights and can even result in drawing the wrong conclusions.

Just having the data in one place doesn't matter if computers or human beings can't make sense of it. That is the power of BDAs. BDAs can help extract the signal from the noise. By doing so, they can give you more confidence in the data you work with, resulting in greater conviction to act on that data, either manually or automatically.

The Big Data Feedback Loop: Acting on Data

The first time you touched a hot stove and got burned, stuck your finger in an electrical outlet and got shocked, or drove over the speed limit and got a ticket, you experienced a feedback loop. Consciously or not, you ran a test, analyzed the result, and acted differently in the future. This is called the Big Data Feedback Loop (see Figure 8-2) and it is the key to building successful BDAs.

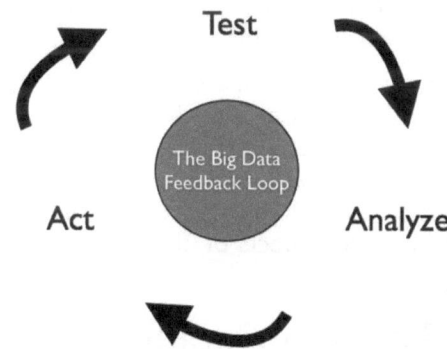

Figure 8-2. The Big Data Feedback Loop

Through testing, you found out that touching a hot stove or getting an electrical shock hurt. You found out that speeding resulted in getting an expensive ticket or getting in a car crash. Or, if you got away with it, you may have concluded that speeding was fun.

Regardless of the outcome, all of these activities gave you feedback. You incorporated this feedback into your data library and changed your future actions based on the data. If you had a fun experience speeding despite the ticket, you might have chosen to do more speeding. If you had a bad experience with a stove, you might have figured out to check if a stove was hot before touching one again in the future.

Such feedback loops are critical when it comes to Big Data. Collecting data and analyzing it isn't enough. You have to be able to reach a set of conclusions from that data and get feedback on those conclusions to determine if they're right or wrong. The more relevant data you can feed into your model and the more times you can get feedback on your hypotheses, the more valuable your insights are.

Historically, running such feedback loops has been slow and time consuming. Car companies, for example, collect sales data and try to draw conclusions about what pricing models or product features cause people to buy more cars. They change prices, revise features, and run the experiment again. The problem is that by the time they conclude their analysis and revise pricing and products, the environment changes. Powerful but gas-guzzling cars were out while fuel-saving cars were in. MySpace lost users while Facebook gained them. And the list goes on.

The benefit of Big Data is that in many cases you can now run the feedback loop much faster. In advertising, BDAs can figure out which ads convert the best by serving up many different ads in real time and then displaying only those that work. They can even do this on a segmented basis—determining which ads convert the best for which groups of people. This kind of A/B testing, displaying different ads to see which perform better, simply could not be done fast enough by humans to work.

Computers can run such tests at massive scale, not only choosing between different ads but actually modifying the ads themselves. They can display different fonts, colors, sizes, and images to figure out which combinations are most effective. This real-time feedback loop, the ability not just to gather massive amounts of data but to test out and act on many different approaches quickly, is one of the most powerful aspects of Big Data.

Achieving Minimum Data Scale

As people move forward with Big Data, it is becoming less a question of gathering and storing the data and more a question of what to do with it. A high-performance feedback loop requires a sufficiently large test set of customers visiting web sites, sales people calling prospects, or consumers viewing ads to be effective.

This test set is called the Minimum Data Scale (MDS; see Figure 8-3). MDS is the minimum amount of data required to run the Big Data Feedback Loop and get meaningful insights from it.

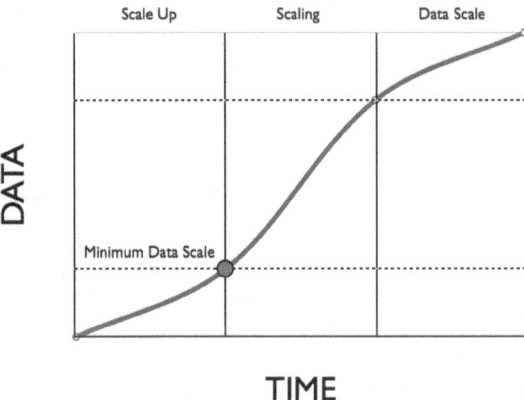

Figure 8-3. The Minimum Data Scale is the minimum amount of data required to derive meaningful conclusions from data

MDS means that a company has enough visitors to its web site, viewers of its advertisements, or sales prospects that it can derive meaningful conclusions and make decisions based on the data. When a company has enough data points to reach MDS, it can use BDAs to tell sales people whom to call next, decide which ad to serve for the highest conversion rate, or recommend the right movie or book.

When that data set becomes so large that it is a source of competitive advantage, it means a company has achieved what early PayPal and LinkedIn analytics guru Mike Greenfield refers to as *Data Scale*.[6] Companies like Amazon, Facebook, Google, PayPal, LinkedIn, and others have all achieved Data Scale.

Creating Big Data Applications

Successful BDAs run part or all of the Big Data Feedback Loop. Some BDAs— powerful analytics and visualization applications, for example—get the data in one place and make it viewable so that humans can decide what to do. Others test new approaches and decide what to do next automatically, as in the case of ad serving or web site optimization.

---

[6]http://numeratechoir.com/2012/05/

The BDAs of today can't help significantly in reaching global maximums. For instance, they can't invent the next iPhone or build the next Facebook. But they can fully optimize the local maximum. They can serve up the right ads, optimize web pages, tell sales people whom to call, and even guide those sales people in what to say during each call.

It is this combination of Data Scale and applications—Big Data Applications— that will fuel tomorrow's billion dollar IPOs.

The Rise of the Big Data Asset: The Heart of a Big Data Startup

The large volume of data that a company collects is the company's Big Data Asset. Companies that use this asset to their advantage will become more valuable than those that don't. They will be able to charge more and pay less, prioritize one prospect over another, convert more customers, and ultimately, retain more customers.

This has two major implications. First, when it comes to startups, there is a massive opportunity to build the applications that make such competitive advantage possible, as out-of-the-box solutions. Enterprises won't have to build these capabilities on their own; they'll get them as applications that are ready to use right away.

Second, companies—both startups and gorillas—that consider data and the ability to act on it a core asset will have a significant competitive advantage over those that don't.

As an example, PayPal and Square are battling it out to disrupt the traditional payments space. The winner will understand its customers better and reach them more efficiently. Both companies have access to massive quantities of transaction data. The one that can act on this data more effectively will come out on top.

Tip You want to build a billion-dollar company? Pay a lot of attention to the creation and nurturing of your Big Data Asset. It is the shiny feature that will attract investors and buyers.

What It Takes to Build a Billion Dollar Company

With all the opportunities in Big Data, what determines the difference between building a small company and building a big one?

As Cameron Myhrvold, founding partner of venture capital firm Ignition Partners, told me, building a billion dollar company doesn't just mean picking a big market. It means picking the right market. Fast food in China is a huge market. But that doesn't make it a good market. It's highly competitive and a tough market to crack.

Building a billion dollar company means riding a big wave. Big Data, cloud, mobile, and social are all clichés for a reason—because they're big waves. Companies that ride those waves are more likely to succeed than the ones that don't. Of course, it's more than that. It's building specific products that align with those waves to deliver must-have value to customers.

Over the past few years, numerous web companies have sprung up. Some of the most successful ones, like Facebook and Twitter, are Big Data companies disguised as consumer companies. They store immense amounts of data. They maintain extensive social graphs that represent how we are connected to each other and to other entities such as brands. Such companies have built-in systems for optimizing what content and which ads they show to each user.

Now there are multiple billion dollar opportunities to bring these kinds of capabilities to enterprises as out-of-the-box solutions in the form of Big Data Applications.

For enterprises, historically, such data sophistication required buying hardware and software and layering custom development on top. Enterprise software products were the basis for such implementations. But they brought with them time-consuming integrations and expensive consulting in order to make sophisticated data analytics capabilities available to end-users. Even then, only experienced analysts could use the systems that companies developed to generate reports, visualizations, and insights.

Now, these kinds of capabilities are available via off-the-shelf BDAs. Operational intelligence for IT infrastructure no longer requires custom scripts—applications from companies like Splunk, a company currently valued at more than $6 billion, make such functionality readily available. Data visualization no longer requires complex programming in languages like R. Instead, business users can create interactive visualizations using products from companies like Tableau Software, which is valued at nearly $5 billion. Both of these Big Data companies just went public in the last few years.

More Big Data companies are gearing to go public soon. Big Data companies AppDynamics and New Relic both process billions of events per month and are expected to file in 2014. Palantir, a data analytics company, recently raised capital at a $9 billion valuation and estimates indicate the company may be doing as much as $1 billion a year in revenue.

Given all the opportunity, perhaps the biggest challenge in reaching a billion for any startup that is able to get meaningful traction is the high likelihood of getting acquired.

When it comes to big vendors in the enterprise space, these days there are a lot of potential acquirers. There are the traditional ultra big, with market caps approaching or more than $100B: Cisco, IBM, Intel, Microsoft, Oracle, and SAP, with a mix of hardware and software offerings.

There are the big players that have an evolving or emerging role in the enterprise, like Amazon, Apple, and Google. There is the $20B–$50B group, like Dell, HP, EMC, Salesforce, and VMWare.

Finally, there are those companies in the sub $10B range (NetApp is just north of that), like BMC, Informatica, Workday, NetSuite, ServiceNow, Software AG, TIBCO, Splunk, and others. This group is a mix of the new, like Workday, ServiceNow, and Splunk, and the older, like BMC and TIBCO. The smaller ones are potential takeover targets for the big. Cisco or IBM might be interested in Splunk, while Oracle or HP might want to buy TIBCO.

Existing public companies that can bulk up by adding more BDAs to their portfolios will do so. Expect Salesforce and Oracle to keep on buying. Salesforce will continue to build out a big enough and comprehensive enough ecosystem that it can't easily be toppled. Oracle will add more cloud offerings so that it can offer Big Data however customers want it, be that on-premise or in the cloud.

Companies like TIBCO and Software AG will keep buying as well. Both companies have been building their cash war chests, likely with an eye toward expanding their portfolios and adding non-organic revenue.

What does that mean for entrepreneurs trying to build the next billion dollar public company? It means they need a great story, great revenue growth, or both. Splunk, for example, started out as a centralized way to view log files. It turned into the Oracle of machine data. That story certainly helped, as did the company's growth.

The small will look to bulk up and the big will look to stay relevant, ramp earnings, and leverage the immense scale of their enterprise sales organizations. That means that entrepreneurs with successful products and revenue growth that want to surpass a billion in market cap will have to work hard not to get taken out too early.

Tip If you are a tech entrepreneur looking for a great exit, don't sell too soon—nor too late. It's a fine line between taking the time to increase your valuation and waiting too long to sell or go public.

Investment Trends: Opportunities Abound

When it comes to Big Data, we are already seeing a host of new applications being built. The BDAs we've seen so far are just the tip of the iceberg. Many are focused on line-of-business issues. But many more will emerge to disrupt entire spaces and industries.

Big enterprises have shown more and more willingness to give smaller vendors a shot—because they need the innovation that these smaller companies bring to the table.

Historically, vendors and customers were both reliant on and at odds with each other. Customers wanted flexibility but couldn't switch due to their heavy custom development and integration investments, while vendors wanted to drive lock-in.

On the other hand, large customers needed comprehensive solutions. Even if they didn't have the best technology in every area, it meant that the CIO or CEO of a big customer could call the CEO of a big vendor and get an answer or an issue resolved.

In a time of crisis such as a system outage, data problems, or other issues, one C could call another C and make things happen. In a world of many complex and interconnected systems—such as flight scheduling, baggage routing, package delivery, and so on—that was and is critical. But nimble startups are bringing a level of innovation and speed to the table that big vendors just can't match. They're creating their services in the cloud using off-the-shelf building blocks. They're getting to market faster, with easy-to-use consumer-like interfaces that larger, traditional vendors struggle to match. As a result, big and small buyers alike are becoming customers of a variety of new Big Data startups that offer functionality and speed that traditional vendors can't match.

There are many opportunities for new applications, both broad and niche. Take the police department of Santa Cruz, California as an example. By analyzing historical arrest records, the police department is able to predict areas where crime will happen. The department can send police officers to areas where crime is likely to happen, which has been shown to reduce crime rates. That is, just having officers in the area at the right time of day or day of week (based on the historical analysis) results in a reduction in crime. The police department of Santa Cruz is assisted by PredPol, a company that is specifically working with this kind of Big Data to make it useful for that specific purpose.

The point is not that investors should back or entrepreneurs should start a hundred predictive policing companies. Rather, it is that, as Myhrvold put it, Big Data is driving the creation of a whole new set of applications. It also means that Big Data isn't just for big companies. If the city of Santa Cruz is being shaped by Big Data, Big Data is going to impact organizations of all sizes as well as our own personal lives, from how we live and love to how we learn. Big Data is no longer only the domain of big companies that have large data analyst and engineering staffs.

The infrastructure is readily available, and as you saw in Chapter 6, much of it is available in the cloud. It's easy to spin up and get started. Lots of public data sets are available to work with. As a result, entrepreneurs will create tons of BDAs. The challenge for entrepreneurs and investors will be to find interesting combinations of data, both public and private, and combine them in specific applications that deliver real value to lots of people over the next few years.

Data from review site Yelp, sentiment data from Twitter, government data about weather patterns—putting these kinds of data sources together could result in some very compelling applications. Banks might better determine who to lend to, while a company picking its next store location might have better insight as to where to locate it.

Big returns, at least when it comes to venture capital, are about riding big waves. As Apple, Facebook, Google, and other big winners have shown, quite often it's not about being first. It's about delivering the best product that rides the wave most successfully.

The other waves investors are betting on are cloud, mobile, and social. Mobile, of course, is disrupting where, when, and how people consume media, interact, and do business. Cloud is making computing and storage resources readily available. And social is changing the way we communicate. Any one of these is a compelling wave to ride.

For entrepreneurs looking to build and investors looking to back the next billion dollar opportunity, it really comes down to three things.

First, it is about choosing the right market—not just a big market. The best markets are ones that have the potential to grow rapidly. Second, it is about riding a wave, be it Big Data, cloud, mobile, or social. Third, as Gus Tai, a general partner at Trinity Ventures, put it, it is about being comfortable with a high level of ambiguity.

If the path is perfectly clear, then the opportunity is, by definition, too small. Clear paths are a lot more comfortable. We can see them. But opportunities that can be the source of a billion new dollars in value are inherently ambiguous. That requires entrepreneurs and investors who are willing to step out of their comfort zones. The next wave of billion dollar Big Data entrepreneurs will be building BDAs where the path is ambiguous, but the goal is crystal clear.

Note   Clear paths to opportunity are not the ones you should be seeking. Others will see them too. Instead, look for opportunities that look foggy or ambiguous from a distance. One area ripe for innovation: making heretofore inaccessible data useful. Think Zillow, AccuWeather, and other companies that make use of large quantities of hard-to-access data.

Big Data "Whitespace" You Can Fill In

Big Data is opening up a number of new areas for entrepreneurship and investment. Products that make data more accessible, that allow analysis and insight development without requiring you to be a statistician, engineer, or data analyst are one major opportunity area. Just as Facebook has made it easier to share photos, new analytics products will make it far easier not just to run analysis but also to share the results with others and learn from such collaborations.

The ability to pull together diverse internal data sources in one place or to combine public and private data sources also opens up new opportunities for product creation and investment. New data combinations could lead to improved credit scoring, better urban planning, and the ability for companies to understand and act on market changes faster than their competition.

There will also be new information and data service businesses. Although lots of data is now on the web—from school performance metrics to weather information to U.S. Census data—lots of this data remains hard to access in its native form.

Real estate web site Zillow consolidated massive amounts of real estate data and made it easy to access. The company went beyond for-sale listings by compiling home sales data that was stored in individual courthouses around the country. Services like Zillow will emerge in other categories. Gathering data, normalizing it, and presenting it in a fashion that makes it easily accessible is difficult. But information services is an area ripe for disruption exactly because accessing such data is so hard.

Innovative data services could also emerge as a result of new data that we generate. Since smartphones come with GPS, motion sensors, and built-in Internet connectivity, they are the perfect option for generating new location-specific data at a low cost.

Developers are already building applications to detect road abnormalities, such as potholes, based on vibrations that phones detect when vehicles move across uneven road surfaces. That is just the first of a host of new BDAs based on collecting data using low-cost sensors built into devices like smartphones.

Big Data Business Models

To get the most value from such whitespace opportunities ultimately requires the financial markets to understand not just Big Data businesses, but subscription business models as well.

Although Splunk is primarily sold via a traditional licensed software plus service model, the company charges based on the amount of data indexed per day. As a result, investors can easily value Splunk on the volume of data companies manage with it. In the case of Splunk, revenue is tied directly to data volume. (Splunk also offers a term license, wherein the company charges a yearly fee.)

For most cloud-based offerings, however, valuations are not so simple. As Tien Tzuo, former Chief Marketing Officer of Salesforce.com and CEO of cloud-based billing company Zuora, points out, financial managers still don't fully appreciate the value of subscription businesses.

This is important because companies will use subscription models to monetize many future BDAs. Tzuo suggests that such models are about building and monetizing recurring customer relationships. That's in contrast to their historical licensed-software and on-premise hardware counterparts, which were "about building and shipping units for discrete one-time transactions."[7] As Tzuo puts it, knowing that someone will give you a $100 once is less valuable than knowing that that person will give you a $100 a year for the next eight years, as they do in subscription businesses.

In the years ahead, companies that have subscription models will be well-positioned to hold onto their customers and maintain steady revenue streams. They won't have to rely on convincing existing customers to upgrade to new software versions in addition to the already hard job of acquiring new customers. Instead they'll simply be able to focus on delivering value.

Apple effectively has the best of both in today's world: repeat one-time purchases in the form of iPads, iPhones, and Macs combined with on-going revenue from the purchase of digital goods like songs and music. That said, even Apple with its incredible brand loyalty may soon need to change to a subscription-based business model. Companies like Spotify are making subscription-based music a viable alternative to one-time purchases.

---

[7]http://allthingsd.com/20121128/wall-street-loves-workday-but-doesnt-understand-subscription-businesses/

Twenty years ago, it was hard for investors to imagine investing exclusively in software businesses. Now such investments are the norm. As a result, it's not hard to envision that despite today's skepticism, 20 years hence most software businesses will be subscription businesses. At a minimum, their pricing will be aligned with actual usage in some way: more data, more cost; less data, less cost.

What such businesses will need to figure out, however, is how to offer cost predictability alongside flexibility and agility. Due to the ease with which customers can spin up more instances or store more data, cloud costs can be wildly unpredictable and therefore difficult to budget for and manage.

A number of different usage-based models exist, particularly for cloud offerings. These include charging based on data volume; number of queries, as in the case of some analytics offerings; or on a subscription basis. Customers who take advantage of such offerings don't need to maintain their own hardware, power, and engineering maintenance resources. More capacity is available when customers need it, less when they don't.

The cost of such flexibility, however, is less predictability around cost. In most cases, such offerings should be less expensive than traditional software. Customers only pay for what they use.

It's easy to run over your mobile phone minutes and get an unexpectedly large bill. Similarly, customers can overuse and over-spend on usage-based software offerings without realizing it.

Such ambiguity has slowed the adoption of some subscription-based services, because CFOs and CIOs rightly demand predictability when it comes to budgeting. Vendors will need to introduce better controls and built-in usage policies so that buyers can more easily manage their expenditures. They'll need to make it easier not just to scale up, but to scale down as well, with little to no engineering effort.

Just over 20 years ago, limited partners—like the managers of university endowments and corporate and municipal pension funds—considering putting capital into venture capital funds thought that investing in software was a radical idea, says Mark Gorenberg, Managing Partner of Hummer Winblad Ventures. Previously they had invested primarily in hardware companies that developed computer chips, networking equipment, and the like. Investors even viewed hardware companies like Cisco that developed a lot of software that ran on such hardware primarily as hardware companies.

The limited partners of the time wondered if there would be any more companies like Microsoft. Now, of course, investing in telecom and hardware is out of favor, and investing in software is the norm. For a long time, in business-to-business (B2B) software, that has meant investing in installed, on-premise software. That's software that companies sell, and then charge for implementing, supporting, and upgrading. It requires vendors to fight an on-going battle of convincing existing customers to upgrade to new versions just to maintain their existing revenue base.

Today's markets still tend to undervalue subscription-based businesses. Although vendors have much work to do in making subscription and data-based pricing models easier to manage and budget, it is not hard to envision a world 20 years from now in which most pricing is subscription-based. That would mean that instead of financial managers discounting subscription-based companies, they would instead discount companies that still sell on a traditional licensed software and service model.

In such a future, those that had adjusted and moved to subscription models aligned with customer usage would be buying up and consolidating those that hadn't, instead of the reverse. If there is one truism of the tech industry, it is that while it takes the gorillas a long, long time to fall, there is always a new, nimble startup that emerges to become the next gorilla. For cynics who claim that such things won't happen, just remember that we have short memories. You need look back no further than at the housing booms and busts or the dot-com crash.

As a result, there will always be another billion dollar company, and the great thing about B2B companies is that there is typically room for a few in each area (unlike with consumer companies like Facebook, which tend to be winner-take-all endeavors). With the waves of Big Data, cloud, mobile, and social setting the stage, the next 20 years of information technology are sure to be even more exciting and valuable than the last 20.

Reach More Customers with Better Data—and Products

How Big Data Creates a Conversation Between Company and Customer

No book on Big Data would be complete without a few words on the Big Data conversation. One of the biggest challenges to widespread adoption of Big Data is the nature of the Big Data conversation itself.

Historically, the discussion around Big Data has been a highly technical one. If the discussion remains highly technical, the benefits of Big Data will remain restricted to those with deep technical expertise. Good technology is critical. But companies must focus on communicating the business value they deliver in order for customers to buy their products and for business leaders to embrace a culture of being data-driven.

In developing the Big Data Landscape, I talked with and evaluated more than 100 vendors, from seed-stage startups to Fortune 500 companies. I spoke with numerous Big Data customers as well.

Many Big Data vendors lead with the technical advantages of their products to the exclusion of talking about business value, or vice versa. A technical-savvy company will often highlight the amount of information its database can store or how many transactions its software can handle per second.

A vision-savvy company, meanwhile, will talk about how it plays in the Big Data space but will lack the concrete technical data to show why its solutions perform better or the specifics about what use cases its product supports and the problems it solves.

Effective communication about Big Data requires both vision and execution. Vision involves telling the story and getting people excited about the possibilities. Execution means delivering on specific business value and having the proof to back it up.

Big Data cannot solve—at least not yet—a lack of clarity about what a product does, who should buy it, or the value a product delivers. Companies that lack clarity on these fronts struggle to sell their products no matter how hard they try.

Thus, there are three key components when it comes to a successful Big Data conversation: vision, value, and execution. "Earth's biggest bookstore," "The ultimate driving machine," and "A developer's best friend," all communicate vision clearly.[1]

But clarity of vision alone is not enough. It must go hand in hand with clear articulation of the value a product provides, what it does, and who, specifically, should buy it.

Based on vision and business value, companies can develop individual stories that will appeal to the customers they're trying so hard to reach as well as to reporters, bloggers, and other members of industry. They can create insightful blog posts, infographics, webinars, case studies, feature comparisons, and all the other marketing materials that go into successful communication—both to get the word out and to support sales teams in explaining their products to customers.

Content, like other forms of marketing, needs to be highly targeted. The same person who cares about teraflops and gigabits may not care as much about which companies in the Fortune 500 use your solution. Both pieces of information are important. They simply matter to different audiences.

---

[1]Amazon.com, BMW, and New Relic, respectively.

Even then, companies can generate a lot of awareness about their products but fail to convert prospects when they land on their web sites. All too often, companies work incredibly hard to get visitors to their sites, only to stumble when it comes to converting those prospects into customers.

Web site designers place buttons in non-optimal locations, give prospects too many choices of possible actions to take, or build sites that lack the information that customers want. It's all too easy to put a lot of friction in between a company and a customer who wants to download or buy.

When it comes to Big Data marketing, it's much less about traditional marketing and much more about creating a conversation that is accessible. By opening up the Big Data conversation, we can all bring the benefits of Big Data to a much broader group of individuals.

Better Marketing with Big Data

Big Data itself can help improve the conversation, especially as more ad spend moves online. In 2013, marketers in the United States spent some $171.7 billion on advertising.[2] As spending on offline channels such as magazines, newspapers, and the yellow pages continues to decline, new ways to communicate with customers online and via mobile keep springing up. Marketers spent $42.8 billion on online advertising in the United States in 2013, and invested some $7.1 billion in mobile advertising.[3]

Google remains the gorilla of online advertising, accounting for some 49.3% of total digital advertising revenue in 2013.[4] Meanwhile, social media such as Facebook, Twitter, and LinkedIn represent not only new marketing channels but new sources of data as well. From a Big Data perspective, the opportunity doesn't stop there.

Marketers have analytics data from visitors to their web sites, customer data from trouble ticketing systems, and actual product usage data. That data that can help them close the loop in understanding how their marketing investments translate into customer action.

---

[2]http://www.emarketer.com/Article/US-Total-Media-Ad-Spend-Inches-Up-Pushed-by-Digital/1010154
[3]http://www.iab.net/about_the_iab/recent_press_releases/press_release_archive/press_release/pr-041014
[4]http://wallstcheatsheet.com/technology/should-google-be-worried-about-the-future-of-online-advertising.html/?a=viewall

Marketing today doesn't just mean spending money on ads. It means that every company has to think and act like a media company. It means not just running advertising campaigns and optimizing search engine listings, but developing content, distributing it, and measuring the results. Big Data Applications can pull the data from all the disparate channels together, analyze it, and make predictions about what to do next—either to help marketers make better decisions or to take action automatically.

■ **Note** Every company now has to think and act like a data-driven media company. Big Data Applications are your new best friend when it comes to understanding all the data streaming into your company and making the wisest long-term decisions possible.

Big Data and the CMO

By 2017, chief marketing officers (CMOs) will spend more on information technology than chief information officers (CIOs), according to industry research firm Gartner.[5] Marketing organizations are making more of their own technology decisions, with less involvement from IT. More and more, marketers are turning to cloud-based offerings to serve their needs. That's because they can try out an offering and discard it if it doesn't perform, without significant up-front cost or time investment.

Historically, marketing expenses have come in three forms: people to run marketing; the costs of creating, running, and measuring marketing campaigns; and the infrastructure required to deliver such campaigns and manage the results.

At companies that make physical products, marketers spend money to create brand awareness and encourage purchasing. Consumers purchase at retail stores, such as car dealerships, movie theaters, and other physical locations, or at online destinations such as Amazon.com.

Marketers at companies that sell technology products often try to drive potential customers directly to their web sites. A technology startup, for example, might buy Google AdWords—the text ads that appear on Google's web site and across Google's network of publishing partners—in the hopes that people will click on those ads and come to their web site. From there, the potential customer might try out the company's offering or enter their contact information in order to download a whitepaper or watch a video, activities which may later result in the customer buying the company's product.

---

[5]http://www.zdnet.com/research-the-devalued-future-of-it-in-a-marketing-world-7000003989/

All of this activity leaves an immense digital trail of information—a trail that is multiplied ten times over, of course, because Google AdWords aren't the only form of advertising companies invest in to drive customers to their web sites. Marketers buy many different kinds of ads across different ad networks and media types. Using the right Big Data tools, they can collect data and analyze the many ways that customers reach them. These range from online chat sessions to phone calls, from web site visits to the product features customers actually use. They can even analyze which segments of individual videos are most popular.

Historically, the systems required to create and manage marketing campaigns, track leads, bill customers, and provide helpdesk capabilities came in the form of expensive and difficult-to-implement installed enterprise software solutions. IT organizations would embark on the time-consuming purchase of hardware, software, and consulting services to get a full suite of systems up and running to support marketing, billing, and customer service operations.

Cloud-based offerings have made it possible to run all of these activities via the Software as a Service (SaaS) model. Instead of having to buy hardware, install software, and then maintain such installations, companies can get the latest and greatest marketing, customer management, billing, and customer service solutions over the web.

Today, a significant amount of the data many companies have on their customers is now in the cloud, including corporate web sites, site analytics, online advertising expenditures, trouble ticketing, and the like. A lot of the content related to company marketing efforts such as press releases, news articles, webinars, slide shows, are now online. Marketers at companies that deliver products such as online collaboration tools or web-based payment systems over the web can now know which content a customer or prospect has viewed, along with demographic and industry information.

The challenge and opportunity for today's marketer is to put the data from all that activity together and make sense of it. For example, a marketer might have their list of customers stored in Salesforce.com, leads from their lead-generation activities stored in Marketo or Eloqua, and analytics that tell them about company web site activity in Adobe Omniture or Google Analytics, a web site analytics product from Google.

Certainly, a marketer could try to pull all that data into a spreadsheet and attempt to run some analysis to determine what's working well and what isn't. But actually understanding the data takes significant analysis. Is a certain press release correlated with more web site visits? Did a particular news article generate more leads? Do visitors to a web site group fit into certain industry segments? What kinds of content appeals to which visitors? Did moving a button to a new location on a web site result in more conversions?

These are all questions that consumer packaged goods (CPG) marketers like Procter & Gamble (P&G) have focused on for years. In 2007, P&G spent $2.62 billion on advertising and in 2010 the company spent $350 million on customer surveys and studies.[6] With the advent of Big Data, the answers are available not just to CPG companies who spend billions on advertising and hundreds of millions on market research each year but also to big and small vendors alike across a range of industries. The promise of Big Data is that today's tech startup can have as much information about its customers and prospects as a big CPG company like P&G.

Another issue for marketers is understanding the value of customers—in particular, how profitable they are. For example, a customer who spends a small amount of money but has lots of support requests is probably unprofitable. Yet correlating trouble ticket data, product usage data, and information about how much revenue a particular customer generated with how much it cost to acquire that customer remains very hard to do.

Big Data Marketing in Action

Although few companies can analyze such vast amounts of data in a cohesive manner today, one thought leader who has been able to perform such analysis is Patrick Moran, vice president of marketing at New Relic. New Relic is an application performance monitoring company. The company makes tools that help developers figure out what's causing web sites to run slowly and make them faster.

Moran has been able to pull together data from systems like Salesforce.com and demand-generation system Marketo along with data from Zendesk, a helpdesk ticketing system, and from Twitter campaigns, on which New Relic spends some $150,000 per month. In conjunction with a data scientist, Moran's team is able to analyze all that data and figure out which Twitter campaigns have the most impact—down to the individual tweets. That helps Moran's team determine which campaigns to spend more on in the future.

The first step in Moran's ability to gather and analyze all that data is having it in the cloud. Just as New Relic itself is a SaaS company, virtually all of the systems from which Moran's team gathers marketing data are cloud-based.

The next step in the process is running a series of marketing campaigns by investing in ads across Google, Twitter, and other online platforms.

---

[6]http://www.valueline.com/Stocks/Highlight.aspx?id=9552 and http://www.chacha.com/question/how-much-does-p%26g-spend-on-advertising

Third, the marketing team gathers all the data from Salesforce.com, Marketo, Twitter campaigns, product usage data, and other forms of data in one place. In New Relic's case they store the data in Hadoop.

Fourth, using the open source statistics package R, the team analyzes the data to determine the key factors that drive the most revenue. For example, they can evaluate the impact on revenue of the customer's geographic location, the number of helpdesk tickets a customer submitted, the path the customer took on New Relic's web site, the tweets a customer saw, the number of contacts a customer has had with a sales rep, and the kind of performance data the customer monitors within the New Relic application. By analyzing all of this data, Moran's team even knows what time of day to run future campaigns. Finally, the team runs a new set of campaigns based on what they've learned.

New Big Data Applications are emerging specifically to make the process that teams like Moran's follow easier. MixPanel, for example, is a web-based application that allows marketers to run segmentation analysis, understand their conversion funnels (from landing page to product purchase), and perform other kinds of marketing analysis.

By aggregating all of this information about customer activity, from ad campaign to trouble ticket to product purchase, it is possible for Big Data marketers to correlate these activities and not only reach more potential customers but to reach them more efficiently.

Marketing Meets the Machine: Automated Marketing

The next logical step in Big Data marketing is not just to bring disparate sources of data together to provide better dashboards and insights for marketers, but to use Big Data to automate marketing. This is tricky, however, because there are two distinct components of marketing: creative and delivery.

The creative component of marketing comes in the form of design and content creation. A computer, for example, can't design the now famous "For everything else there's MasterCard" campaign. But it can determine whether showing users a red button or a green button, a 12-point font or a 14-point font, results in more conversions. It can figure out, given a set of potential advertisements to run, which ones are most effective.

Given the right data, a computer can even optimize specific elements of a text or graphical ad for a particular person. For example, an ad optimization system could personalize a travel ad to include the name of the viewer's city: "Find the lowest fares between San Francisco and New York" instead of just

"Find the lowest fares."[7] It can then determine whether including such information increases conversion rates.

■ **Note** Increasingly, computers will make many minor marketing decisions—which ad to show, what color the background should be, where to place the "Tell Me More" buttons, and so forth. They will have to, given the scale of today's Big Data marketing machine. No human, or team of humans, can make as many effective decisions as fast.

In theory, human beings could perform such customizations manually, and in the past, they did. Graphic artists used to—and some still do—customize each ad individually. Web developers would set up a few different versions of a web page and see which one did the best. The problem with these approaches is two-fold. They're very limited in the number of different layouts, colors, and structures a marketer can try. There's also no easy way to customize what is shown to each individual. A different button location, for example, might work better for one group of potential customers but not for another. Without personalization, what results in higher conversion rates for one group of customers could results in lower conversion rates for another.

What's more, it's virtually impossible to perform such customizations for thousands, millions, or billions of people by hand. And that is the scale at which online marketing operates. Google, for example, serves an average of nearly 30 billion ad impressions per day.[8] That's where Big Data systems excel: when there is a huge volume of data to deal with and such data must be processed and acted upon quickly.

Some solutions are emerging that perform automated modeling of customer behavior to deliver personalized ads. TellApart and AdRoll offer retargeting applications. They combine automated analysis of customer data with the ability to display relevant advertisements based on that data. TellApart, which recently hit a $100 annual run rate, identifies shoppers that have left a retailer's web site and delivers personalized ads to them when they visit other web sites, based on the interests that a given shopper showed while browsing the retailer's site.[9] This kind of personalized advertising brings shoppers back to

---

[7]http://www.adopsinsider.com/ad-ops-basics/dynamic-creative-optimization-where-online-data-meets-advertising-creative/
[8]http://www.business2community.com/online-marketing/how-many-ads-does-google-serve-in-a-day-0322253
[9]http://techcrunch.com/2013/12/12/ad-tech-startup-tellapart-hits-100-million-revenue-run-rate/

the retailer's site, often resulting in a purchase. By analyzing shopper behavior, TellApart is able to target high-quality customer prospects while avoiding those who aren't ultimately likely to make a purchase.

When it comes to marketing, automated systems are primarily involved in large-scale ad serving and in lead-scoring, that is, rating a potential customer lead based on a variety of pre-determined factors such as the source of the lead. These activities lend themselves well to data mining and automation. They are well-defined processes with specific decisions that need to be made, such as determining whether a lead is good, and actions that can be fully automated, such as choosing which ad to serve.

Plenty of data is available to help marketers and marketing systems optimize content creation and delivery. The challenge is putting it to work.

Social media scientist Dan Zarrella has studied millions of tweets, likes, and shares, and has produced quantitative research on what words are associated with the most re-tweets, the optimal time of day to blog, and the relative importance of photos, text, video, and links.[10] The next step in Big Data meets the machine will be Big Data Applications that combine research like Zarrella's with automated content campaign management.

In the years ahead, you'll see intelligent systems continue to take on more and more aspects of marketing. These systems won't just score leads, they'll also determine which campaigns to run and when to run them. They'll customize web sites so that the ideal site is displayed to each individual visitor. Marketing software won't just be about dashboards that help humans make better decisions—useful as that is. With Big Data, marketing software will be able to run campaigns and optimize the results automatically.

The Big Data Content Engine

When it comes to creating content for marketing, there are really two distinct kinds of content most companies need to create: high-volume and high-value. Amazon, for example, has some 248 million pages stored in Google's search index.[11] Such pages are known as the long tail. People don't come across any individual page all that often, but when someone is searching for a particular item, the corresponding web page is there in Google's index to be found. Consumers searching for products are highly likely to come across an Amazon page while performing their search.

---

[10]http://danzarrella.com
[11]http://blog.hubspot.com/blog/tabid/6307/bid/21729/3-Marketing-Lessons-From-Amazon-s-Web-Strategy.aspx

Human beings can't create each of those pages. Instead, Amazon automatically generates its pages from its millions of product listings. The company create pages that describe individual products as well as pages that are amalgamations of multiple products: there's a headphones page, for example, that lists all the different kinds of headphones along with individual headphones and text about headphones in general. Each page, of course, can be tested and optimized.

Amazon has the advantage not only of having a huge inventory of products—its own and those listed by merchants that partner with Amazon—but a rich repository of user-generated content, in the form of product reviews, as well. Amazon combines a huge Big Data source, its product catalog, with a large quantity of user-generated content.

This makes Amazon not only a leading product seller but also a leading source of great content. In addition to reviews, Amazon has product videos, photos (both Amazon and user-supplied), and other forms of content. Amazon reaps the rewards of this in two ways: it is likely to be found in search engine results and users come to think of Amazon as having great editorial content. Instead of just being a destination where consumers go to buy, Amazon becomes a place where consumers go to do product research, making them more likely to make a purchase on the site.

Other companies, particularly e-commerce companies with large existing online product catalogs, have turned to solutions like BloomReach. BloomReach works with web sites to generate pages for the search terms that shoppers are looking for. For example, while an e-tailer might identify a product as a kettle, a shopper might search for the term "hot pot." The BloomReach solution ensures that sites display relevant results to shoppers, regardless of the exact term the shopper searches for.

Amazon isn't the only company that wouldn't traditionally be considered a media company that has turned itself into exactly that. Business networking site LinkedIn has too. In a very short time, *LinkedIn Today* has become a powerful new marketing channel. It has transformed the business social networking site into an authoritative source of content and delivered a valuable service to the site's users in the process.

LinkedIn used to be a site that users would occasionally frequent when they wanted to connect with someone or they were starting a new job search. *LinkedIn Today* has made the site relevant on a daily basis by curating relevant news from around the web and updates from users of the site itself.

LinkedIn goes a step further than most traditional media sites by showing users content that is relevant to them based on their interests and their network. The site brings users back via a daily email that contains previews of the latest news. LinkedIn has created a Big Data content engine that drives new traffic, keeps existing users coming back, and maintains high levels of engagement on the site.

How can a company that doesn't have millions of users or product listings create content at the Big Data scale? I'll answer that question in a moment. But first, a few words on marketing and buying Big Data products.

The New PR: Big Data and Content Marketing

When it comes to driving demand for your products and keeping prospects engaged, it's all about content creation: blog posts, infographics, videos, podcasts, slide decks, webinars, case studies, emails, newsletters, and other materials are the fuel that keep the content engine running.

Since 1980, the number of journalists "has fallen drastically while public relations people have multiplied at an even faster rate."[12] In 1980, there were .45 public relations (PR) workers per 100,000 people compared to .36 journalists. In 2008 there were twice as many PR workers, .90 for every 100,000 people, compared with .25 journalists. That means there are more than three PR people for every journalist, which makes getting your story covered by a reporter harder than ever before. Companies, Big Data and otherwise, have to create useful and relevant content themselves to compete at the Big Data scale.

In many ways, content marketing is the new advertising. As of 2011, according to NM Incite, a Nielsen/McKinsey company, there were some 181 million blogs worldwide compared to only 36 million in 2006.[13] But the good news for companies trying to get the word out about their products is that many of these blogs are consumer-oriented with small audiences, and creating a steady stream of high-quality content is difficult and time-consuming. A lot more people consume content than create it. A study by Yahoo research[14] showed that about 20,000 Twitter users (just .05% of the user base) generated 50% of all tweets.[15]

Content marketing means putting as much effort into marketing your product as you put into marketing the content you create about your product. Building great content no longer means simply developing case studies or product brochures specifically about your product but delivering news stories, educational materials, and entertainment.

---

[12]http://www.propublica.org/article/pr-industry-fills-vacuum-left-by-shrinking-newsrooms/single
[13]http://blog.nielsen.com/nielsenwire/online_mobile/buzz-in-the-blogosphere-millions-more-bloggers-and-blog-readers/
[14]http://research.yahoo.com/pub/3386
[15]http://blogs.hbr.org/cs/2012/12/if_youre_serious_about_ideas_g.html

In terms of education, IBM for example, has an entire portfolio of online courses. Vacation rental site Airbnb created Airbnb TV to showcase its properties in cities around the world, which in the process showcased Airbnb itself. You can no longer just market your product; you have to market your content too, and that content has to be compelling in its own right.

■ **Note** Content marketing will be a big part of your future. Gaining market adoption isn't just about developing great products—it's about ensuring people understand the value of those products. Fortunately, that's where content marketing comes to the rescue.

Crowdsource Your Way to Big Data Scale

Producing all that content might seem like a daunting and expensive task. It needn't be. Crowdsourcing, which involves outsourcing tasks to a distributed group of people, is the easy way to generate that form of unstructured data that is so critical for marketing: content.[16]

Many companies already use crowdsourcing to generate articles for search engine optimization (SEO), articles that help them get listed and ranked more highly in sites like Google. Many people associate such content crowdsourcing with high-volume, low-value forms of content. But today it is possible to crowdsource high-value, high-volume content as well.

Crowdsourcing does not replace in-house content development. But it can augment it. A wide variety of sites now provide crowdsourcing services. Amazon Mechanical Turk (AMT) is frequently used for tasks like content categorization and content filtering, which are difficult for computers but easy for humans. Amazon itself uses AMT to determine if product descriptions match their images. Other companies build on top of the programming interfaces that AMT supports to deliver vertical-specific services such as audio and video transcription.[17]

Sites like Freelancer.com and oDesk.com are frequently used to find software engineers or to create large volumes of low-cost articles for SEO purposes, while sites like 99designs and Behance make it possible for creative professionals, such as graphic designers, to showcase their work and for content buyers to line up designers to deliver creative work. Meanwhile, TaskRabbit is applying crowdsourcing to offline tasks such as food delivery, shopping, house cleaning, and pet sitting.

---

[16]http://en.wikipedia.org/wiki/Crowdsourcing
[17]Speechpad.com, which the author co-founded, is one such example.

One of the primary differences between relatively low-value content created exclusively for SEO purposes—which despite (or perhaps because of) its goal is having progressively less impact on search results—and high-value content is the authoritative nature of the latter. Low-value content tends to provide short-term fodder for search engines in the form of an article written to catch a particular keyword search.

High-value content, in contrast, tends to read or display more like professional news, education, or entertainment content. Blog posts, case studies, thought leadership pieces, technical writeups, infographics, video interviews, and the like fall into this category. This kind of content is also the kind that people want to share. Moreover, if your audience knows that you have interesting and fresh content, that gives them more reason to come back to your site on a frequent basis and a higher likelihood of staying engaged with you and your products.

The key to such content is that it must be newsworthy, educational, entertaining, or better yet, a combination of all three. The good news for companies struggling to deliver this kind of content is that crowdsourcing now makes it easier than ever.

Crowdsourcing can come in the form of using a web site like 99designs, but it doesn't have to. As long as you provide a framework for content delivery, you can plug crowdsourcing in to generate the content. For example, if you create a blog for your web site, you can author your own blog posts but also publish those authored by contributors, such as customers and industry experts.

If you create a TV section of your site, you can post videos that are a mix of videos you create yourself, videos embedded from other sites, such as YouTube, and videos produced through crowdsourcing. Those producers can be your own employees, contractors, or industry experts conducting their own interviews. You can crowdsource webinars and webcasts in much the same way. Simply look for people who have contributed content to other sites and contact them to see if they're interested in participating on your site.

Using crowdsourcing is an efficient way to keep your high-value content production machine humming. It simply requires a content curator or a content manager to manage the process. Of course, even that can be crowdsourced. Most importantly, as it relates to Big Data, as you create your content, you can use analytics to determine which content is most appealing, interesting, and engaging for your users. By making Big Data an integral part of your content marketing strategy, you can bring together the best of both worlds—rich content with leading edge analytics that determine which content is a hit and which isn't.

Every Company Is Now a Media Company

In addition to creating content that's useful in the context of your own web site, it's also critical to create content that others will want to share and that bloggers and news outlets will want to write about. That means putting together complete content packages. Just as you would include an image or video along with a blog post on your own site, you should do the same when creating content you intend to pitch to others.

Some online writers are now measured and compensated based on the number of times their posts are viewed. As a result, the easier you make it for them to publish your content and the more compelling content you can offer them, the better. For example, a press release that comes with links to graphics that could potentially be used alongside an article is easier for a writer to publish than one that doesn't.

A post that is ready to go, in the form of a guest post, for example, is easier for an editor or producer to work with than a press release. An infographic that comes with some text describing what its key conclusions are is easier to digest than a graphic by itself.

Once your content is published, generating visibility for it is key. Simply announcing a product update is no longer sufficient. High-volume, high-quality content production requires a media company-like mindset. Crowdsourcing is still in its infancy but you can expect the market for it to continue to grow in the coming years.

Measure Your Results

On the other end of the spectrum from content creation is analyzing all that unstructured content to understand it. Computers use natural language processing and machine learning algorithms to understand unstructured text, such as the half billion tweets that Twitter processes every day. This kind of Big Data analysis is referred to as sentiment analysis or opinion mining.

By evaluating posts on Internet forums, tweets, and other forms of text that people post online, computers can determine whether consumers view brands positively or negatively. Companies like Radian6, which Salesforce.com acquired for $326 million in 2010, and Collective Intellect, which Oracle acquired in 2012, perform this kind of analysis. Marketers can now measure overall performance of their brand and individual campaign performance.

Yet despite the rapid adoption of digital media for marketing purposes, measuring the return on investment (ROI) from marketing remains a surprisingly inexact science. According to a survey of 243 CMOs and other executives,

57% of marketers don't base their budgets on ROI measures.[18] Some 68% of respondents said they base their budgets on historical spending levels, 28% said they rely on gut instinct, and 7% said their marketing spending decisions weren't based on any metrics.

The most advanced marketers will put the power of Big Data to work, removing more unmeasurable components from their marketing efforts and continuing to make their marketing efforts more data-driven, while others continue to rely on traditional metrics such as brand awareness or no measurement at all. This will mean a widening gap between the marketing haves and the marketing have-nots.

While marketing at its core will remain creative, the best marketers will use tools to optimize every email they send, every blog post they write, and every video they produce. Ultimately every part of marketing that can be done better by an algorithm—such as choosing the right subject line or time of day to send an email or publish a post—will be. Just as so much trading on Wall Street is now done by quants, large portions of marketing will be automated in the same way.[19] Creative will pick the overall strategy. Quants will run the execution.

Of course, great marketing is no substitute for great product. Big Data can help you reach prospective customers more efficiently. It can help you better understand who your customers are and how much they're spending. It can optimize your web site so those prospects are more likely to convert into customers once you've got their attention. It can get the conversation going. But in an era of millions of reviews and news that spreads like wildfire, great marketing alone isn't enough. Delivering a great product is still job one.

---

[18]http://adage.com/article/cmo-strategy/study-finds-marketers-practice-roi-preach/233243/

[19]http://www.b2bmarketinginsider.com/strategy/real-time-marketing-trading-room-floor

How Big Data Is Changing the Way We Live

Sectors Ripe for Big Data Projects

The starting line at Ironman France 2012 was eerily quiet. There was a nervous tension in the air as 2,500 people got ready to enter the water and spend as many as the next 16 hours trying to complete what was for some the goal of a lifetime. Made famous by the Ironman World Championship held in Kona, Hawaii every year, an event that started out as a 15-person race in 1978 is now a global phenomenon.

Ironman contenders, like almost all athletes, are some of the most data-driven people on earth. Completing the race, which consists of a 2.4-mile swim, a 112-mile bike, and a 26.2-mile marathon, takes focus, perseverance, and training.

It also requires an incredible amount of energy. Ironman athletes burn some 8,000 to 10,000 calories during the race.[1] To put that in perspective, human beings burn approximately 2,000 to 2,500 calories on an average day. Often called the fourth sport of triathlon, nutrition can mean the difference between finishing and bonking, which is athlete-speak for running out of energy.

---

[1]http://www.livestrong.com/article/232980-the-calories-burned-during-the-ironman-triathlon/

As a result, both preparing for an Ironman and finishing the event requires incredible attention to data. Athletes that don't put in enough miles won't have enough endurance to finish come race day. And even those who have trained won't cross the finish line if they don't take in enough calories and water to keep their bodies moving.

In the fall of 2011, I decided to train for a full Ironman. Over the course of the next nine months, I learned more about training and nutrition and gathered more data about my personal fitness and health than I ever had before. I would regularly upload my training data to a web site called Garmin Connect, developed by the well-known maker of GPS devices.

Remarkably, as of May 2014, athletes had logged more than 5 billion miles of user activity (stored in a 40 terabyte database) on the Garmin Connect web site. They weren't just logging miles. When it comes to training and events, they were also logging elevation gain and loss, speed, revolutions per minute on their bikes, calories, and heart rate. Off the course, they were uploading metrics on their weight, body fat percentage, body water percentage, muscle mass, and daily calorie intake, among other health measurements.

One might think that capturing, storing, and analyzing such an immense amount of data would cost thousands if not tens of thousands of dollars or more. But watches with built-in GPS are now available for under $100 and scales that measure body composition are available for just over that. Measurement devices come in all forms and there are easy-to-use, free and low-cost measurement and logging applications for iPhone and Android devices. What's more, the Garmin Connect service and others like it are free.

This combination of low-cost devices and applications for capturing a wide variety of data, combined with the ability to store and analyze large volumes of data inexpensively, is an excellent example of the power of Big Data. It shows how Big Data isn't just for large enterprises, but for all of us. It's something that can help us in our everyday lives. And it points the way, as we shall see, toward many areas ripe for product development.

Personal Health Applications

Taking the capture and analysis of our personal health information one step further by applying Big Data to personal genetics is DNA testing and data analytics company 23andMe.[2] Since its founding in 2006 by Anne Wojcicki, the company's CEO and wife of Google co-founder Sergey Brin, the company has analyzed the saliva of more than 400,000 people.

---

[2]https://www.23andme.com/

By analyzing genomic data, the company identifies individual genetic disorders, such as Parkinson's, as well as genetic propensities such as obesity.[3] By amassing and analyzing a huge database of personal genetic information, the company hopes not only to identify individual genetic risk factors that may help people improve their health and live longer, but more general trends as well.

As a result of its analysis, the company has identified some 180 previously unknown traits, including one called the "photic sneeze reflex," which refers to the tendency to sneeze when moving from darkness to bright sunlight, as well as another trait associated with people's like or dislike of the herb cilantro.[4]

Using genomic data to provide insights for better healthcare is in reality the next logical step in an effort first begun in 1990. The Human Genome Project's (HGP) goal was to map all of the approximately 23,000 genes that were ultimately found to make up our DNA. The project took 13 years and $3.8 billion in funding to complete.

Remarkably, storing extensive human genome data doesn't need to take up that much physical space. According to one analysis, human genes can be stored in as little as 20 megabytes—consuming about the same amount of space as a handful of songs stored on your iPod.

How is that possible? About 99.5% of the DNA for any two randomly selected people is exactly the same.[5] Thus, by referring to a reference sequence of the human genome, it's possible to store just the information needed to turn the reference sequence into one that is specific to any one of us.

Although the DNA information of any individual takes up a lot of space in its originally sequenced form—a set of images of DNA fragments captured by a high-resolution camera—once those images are turned into the As, Cs, Gs, and Ts that make up our DNA, the sequence of any particular person can be stored in a highly efficient manner.

Given the sequence of any one person alone, it is hard to produce any informative conclusions. To gain real insight, we must take that data and combine it with scientific research and other forms of diagnosis as well as with changes in behavior or treatments to realize an impact on our health.

---

[3]http://www.theverge.com/2012/12/12/3759198/23andme-genetics-testing-50-million-data-mining
[4]http://blog.23andme.com/health-traits/sneezing-on-summer-solstice/
[5]http://www.genetic-future.com/2008/06/how-much-data-is-human-genome-it.html

From this is should be clear that it is not always the ultimate size of the data that makes it Big Data. The ability not just to capture data but to analyze it in a cost-effective manner is what really makes Big Data powerful. While the original sequencing of the human genome cost some $3.8 billion, today you can get an analysis of your own DNA for $99 from 23andMe. Industry experts believe that that price is subsidized and that the actual cost of individual DNA analysis is more in the $500 to $1000 range. But even so, in just under a decade the cost of sequencing has dropped by multiple orders of magnitude. Just imagine what will happen in the next decade. In the long-run, it is likely that such companies hope not just to offer DNA analysis services, but to offer products and treatments customized to your personal profile as well or to work with pharmaceutical companies and doctors to make such personalized treatments possible. Entrepreneurial opportunities abound in finding new applications of previously unavailable data like DNA information.

Another company, Fitbit, has the goal of making it easier to stay healthy by making it fun. The company sells a small device that tracks your physical activity over the course of the day and while you sleep. Fitbit also offers a free iPhone app that lets users log food and liquid intake.

By tracking their activity levels and nutrition intake, users can figure out what's working well for them and what's not. Nutritionists advise that keeping an accurate record of what we eat and how much activity we engage in is one of the biggest factors in our ability to control our weight, because it makes us accountable.

Fitbit is collecting an enormous amount of information on people's health and personal habits. By doing so, it can show its users helpful charts to help them visualize their nutrition and activity levels and make recommendations about areas for improvement.

■ **Tip** Many health-related products that have historically required expensive, proprietary hardware and software can be redeveloped as consumer-friendly applications that run on a smartphone. Now the latest medical technology can be made available to a much broader group of doctors and patients than ever before.

Another device, the BodyMedia armband, captures over 5,000 data points every minute, including information about temperature, sweat, steps, calories burned, and sleep quality.[6] The armband has been featured on NBC's *The Biggest Loser*, a reality game show focused on weight loss.

---

[6]http://www.bodymedia.com/Professionals/Health-Professionals

Strava combines real-world activity data with virtual competition by taking such challenges outdoors. The company's running and cycling application for iPhone and Android devices is specifically designed to take advantage of the competitive natures of sporting activities. Fitness buffs can compete for leader board spots on a diverse set of real-world segments, such as cycling from the bottom to the top of a challenging hill, and compare their results on Strava's web site. The company also offers pace, power, and heart rate analysis to help athletes improve.

According to an American Heart Association article entitled *The Price of Inactivity*, 65% of all adults are obese or overweight.[7] Sedentary jobs have increased 83% since 1950 and physically active jobs now make up only about 25% of the workforce. Americans work an average of 47 hours per week, 164 more hours per year than they did 20 years ago. Obesity costs American companies an estimated $225.8 billion per year in health-related productivity losses. As a result, devices like the Fitbit and the Nike FuelBand stand to make a real impact on rising healthcare costs and personal health.

One iPhone app can even check your heart rate by reading your face or detecting the pulse rate in your finger. Biofeedback app company Azumio has had more than 20 million downloads of its mobile applications, which can do everything from measure your heart rate to detect your stress level. Although Azumio started out developing individual applications, over time it will be able to measure data across millions of users and provide them with health insights.

There is an opportunity when it comes to Big Data health applications to apply some of the same approaches Facebook and others use for online advertising to improve health. Facebook figures out which advertisements produce the most conversions for users who are similar and optimizes the advertisements it shows as a result. Similarly, future Big Data health applications could use data collected from millions of users not just to monitor health, but to make suggestions for improvement—based on what worked for others with similar profiles.

■ **Tip** Future Big Data health applications could use data from millions of users not just to monitor health, but to make suggestions for improvement—based on what worked for others with similar profiles.

Azumio has already introduced a fitness application called Fitness Buddy, a mobile fitness app with more than 1,000 exercises, 3,000 images and animations, and an integrated fitness journal. Meanwhile, the company's Sleepy Time application monitors sleep cycles using an iPhone. Such applications present intriguing possibilities for Big Data and health and are a lot more convenient and less expensive than the equipment used in traditional sleep labs.

---

[7]http://www.heart.org/HEARTORG/GettingHealthy/PhysicalActivity/StartWalking/The-Price-of-Inactivity_UCM_307974_Article.jsp

Data collected by such applications can tell us what's going on in the moment as well as offer us a picture of our health over time. If our resting heart rate is fluctuating, that may indicate a change in our health status, for example. By working with health data collected across millions of people, scientists can develop better algorithms for predicting future health. Applications can make better suggestions about changes we should make to improve our health.

One of the most compelling aspects of such applications are the ways in which they make it easier to monitor health information over time. Historically, such data collection required specialized and inconvenient devices or a trip to the doctor's office. Inconvenience and expense made it difficult for most people to track basic health information. With Big Data, data collection and analysis becomes much easier and more cost-effective. In one example, Intel and the Michael J. Fox Foundation are working together on a project combining wearable devices and Big Data to gain better insight into Parkinson's disease. Patients use a device like the FitBit, which collects data about them throughout the day. Historically, doctors have collected data about Parkinson's patients only during medical exams. Because symptoms can vary from one minute to the next, such exams may not provide an accurate picture of a patient's health. By combining low-cost devices with data collection, researchers will now be able to analyze "patient gait, tremors and sleep patterns, among other metrics."[8]

The availability of low-cost personal health monitoring applications and related technologies has even spawned an entire movement in personal health. Quantified Self is "a collaboration of users and tool makers who share an interest in self-knowledge through self-tracking."[9] The founders of the Quantified Self movement are two former editors of *Wired* magazine, Kevin Kelly and Gary Wolf. Wolf is known for his TED talk,[10] "The Quantified Self," in which he highlights all of the data we can collect about ourselves, and his *New York Times* article, "The Data Driven Life."[11]

New applications show just how much heath data can be collected via inexpensive devices or devices like smartphones that we already have combined with the right software applications. Put that data collection ability together with low-cost cloud services for analysis and visualization and the area of personal health and Big Data has significant potential to improve health and reduce healthcare costs.

---

[8]http://www.usatoday.com/story/news/nation/2014/08/13/michael-j-fox-parkinsons-intel/13719811/

[9]http://quantifiedself.com/about/

[10]http://www.ted.com/talks/gary_wolf_the_quantified_self.html

[11]http://www.vanityfair.com/culture/2013/02/quantified-self-hive-mind-weight-watchers and http://www.nytimes.com/2010/05/02/magazine/02self-measurement-t.html

---

■ **Tip**   The intersection of health and fitness, body sensors, and analysis technology residing in the cloud presents a wealth of opportunity for application development.

Big Data and the Doctor

Of course, even with such applications, there are times when we need to go to the doctor. A lot of medical information is still collected with pen and paper, which has the benefit of being easy to use and low-cost. But this also result in errors when it comes to recording patient information and billing. Having paper-based records spread across multiple locations also makes it challenging for healthcare providers to access critical information about a patient's health history.

The HITECH—Health Information Technology for Economic and Clinical Health—Act was enacted in 2009 to promote the use of health information technology, and in particular, the adoption of Electronic Health Records or EHRs. It offers healthcare providers financial incentives to adopt EHRs through 2015 and provides penalties for those who don't adopt EHRs after that date. Electronic Medical Records (EMRs) are digital versions of the paper records that many physicians use today. In contrast, an EHR is intended as a common record of a patient's health that can be easily accessed by multiple healthcare providers.[12] The HITECH Act and the need for both EMRs and EHRs is driving the digitization of a lot of healthcare data, and opening up new avenues for analysis as a result.

New applications like drchrono allow physicians to capture patient information using iPads, iPhones, Android devices, or web browsers. In addition to capturing the kind of patient information previously recorded using pencil and paper, doctors get integrated speech-to-text for dictation, the ability to capture photos and videos, and other features.

EHRs, DNA testing, and newer imaging technologies are generating huge amounts of data. Capturing and storing such data presents a challenge for healthcare providers but also an opportunity. In contrast to historically closed hospital IT systems, newer, more open systems combined with digitized patient information could provide insights that lead to medical breakthroughs.

IBM's Watson computer became famous for winning *Jeopardy!* The Memorial Sloan Kettering Cancer Center is now using *Watson* to develop better decision support systems for cancer treatment.[13] By analyzing data from EHRs and academic research, the hope is that Watson will be able to provide doctors with better information for making decisions about cancer treatments.

---

[12]http://www.healthit.gov/buzz-blog/electronic-health-and-medical-records/emr-vs-ehr-difference/
[13]http://healthstartup.eu/2012/05/top-big-data-opportunities-for-health-startups/

Such analysis can lead to additional insights as well. For example, an intelligent system can alert a doctor to other treatments and procedures normally associated with those she's recommending. These systems can also provide busy doctors with more up-to-date information on the latest research in a particular area. The right medical analytics software can even offer personalized recommendations based on data about other patients with similar health profiles.

■ **Tip** Knowledge workers in the enterprise have always benefited from better access to data. Now the opportunity exists to bring the most up-to-date information to healthcare providers and to enable them to make recommendations based on your personal health profile.

The amount of data that all these systems capture and store is staggering. More and more patient data will be stored digitally, and not just the kind of information that we provide on health questionnaires or that doctors record on charts. Such information also includes digital images from devices like iPhones and iPads and from newer medical imaging systems, such as x-ray machines and ultrasound devices, which now produce high-resolution digital images.

In terms of Big Data, that means better and more efficient patient care in the future, the ability to do more self-monitoring and preventive health maintenance, and, of course, a lot more data to work with. One key challenge is to make sure that data isn't just collected for the sake of collecting it but that it can provide key insights to both healthcare providers and individuals. Another key challenge will be ensuring patient privacy and confidentiality as more data is collected digitally and used for research and analysis.

Big Data and Health Cures

A few years ago, I got a rather strange email from my dad. Holding a doctorate in chemistry, my Dad thrives on data. He had had some tests done that showed that his PSA levels, which I would later find out meant Prostate-Specific Antigen, were significantly above normal.

Higher PSA levels are highly correlated with prostate cancer. This raised two key questions. The first was whether my dad actually had cancer. The test did not reveal cancer cells. Rather, higher levels of PSA were often found in people that ultimately were diagnosed with prostate cancer. The difficulty is that not all people with higher PSA levels have cancer. Some of them just have higher PSA levels.

The second challenge my Dad faced was what to do with the information. His options were at the same time simple and complex. On the one hand, he could do nothing. As my personal doctor told me in his classically objective manner, "it's usually something else that kills them first." However, my Dad would have to live with the psychological impact of having a slowly worsening disease, which ultimately, if it did spread, he would likely be too old to do anything about.

On the other hand, he could take action. Action would come in the form of a range of treatments, from hormone therapy to ablative surgery to the complete removal of his prostate. But the treatment might prove worse than the cure.

"What should I do?" my Dad asked the doctor. The doctor gave him the only answer he could: "It's up to you. It's your life."

In the case of hormone therapy, which he ultimately chose, my Dad suffered depression, cold sweats, and extensive periods of difficulty sleeping. Had he chosen the surgery he would have been looking at a year or more of a colostomy bag. A few months later a research study was published indicating that the best treatment for prostate cancer may be not to test for it at all. Apparently the microscopic hole associated with the tests can allow the cancer, which is contained in the prostate gland, to escape.

Therein lie two important lessons about our use of data.

First, data can give us greater insights. It can deliver more relevant experiences. It can allow computers to predict what movie we'll want to watch or what book we'll want to buy next. But when it comes to things like medical treatment, the decisions about what to do with those insights aren't always obvious.

Second, our insights from data can evolve. Insights from data are based on the best data we have available at the time. Just as fraud detections systems try to identify fraudsters based on pattern recognition, those systems can be improved with better algorithms based on more data. Similarly, the suggested approaches to different medical conditions change as we get more data.

In men, the cancers with the highest mortality rates are lung, prostate, liver, and colorectal cancer, while in women the cancers that strike highest are lung, breast, and colorectal cancer. Smoking, a leading cause of lung cancer, has dropped from a rate of 45% of the U.S. population in 1946 to 25% in 1993 to 18.1% as of 2012.[14] However, the five-year survival rate for those with lung cancer is only 15.5%, a figure that hasn't changed in 40 years.[15]

Despite then President Richard Nixon declaring a national war on cancer in 1971, there remains no universal prevention or cure for cancer. That's in large part because cancer is really hundreds of diseases, not just one. There are more than 200 different types of cancer.[16]

---

[14]http://www.cdc.gov/tobacco/data_statistics/fact_sheets/fast_facts/
[15]http://www.lungcancerfoundation.org/who-we-are/the-right-woman-for-the-job/
[16]http://www.cancerresearchuk.org/cancer-help/about-cancer/cancer-questions/how-many-different-types-of-cancer-are-there

The National Cancer Institute (NCI), which is part of the National Institutes of Health, has a budget of about $5 billion per year for cancer research.[17] Some of the biggest advances in cancer research have been the development of tests to detect certain types of cancer, such as a simple blood test to predict colon cancer, which was discovered in 2004.

Other advances have been those linking cancer to certain causes, such as a study in 1954 that first showed a link between smoking and lung cancer, and a study in 1955 that showed that the male hormone testosterone drives the growth of prostate cancer while the female hormone estrogen drives the growth of breast cancer. Still further advances have come in the approaches to treating cancer: the discovery, for example, of dendritic cells, which became the basis for cancer vaccines and the discovery of angiogenesis, the process by which tumors create a network of blood vessels to bring them the oxygen that allows them to grow.[18]

More recently, Big Data has been playing a bigger role. The National Cancer Institute's CellMiner, for example, is a web-based tool that gives researchers access to large quantities of genomic expression and chemical compound data. Such technology makes cancer research more efficient. In the past, working with such data sets often meant dealing with unwieldy databases that made it difficult to analyze and integrate data.[19]

Historically, there was a big gap between the people who wanted to answer questions by using such data and those who had access to the data. Technologies like CellMiner make that gap smaller. Researchers used CellMiner's predecessor, a program called COMPARE, to identify a drug with anticancer activity, which turned out to be helpful in treating some cases of lymphoma. Those researchers are now using CellMiner to figure out biomarkers that will tell them which patients are likely to respond favorably to the treatment.

One of the biggest impacts the researchers cite is the ability to access data they couldn't easily get to before. That is a critical lesson not just for cancer researchers but for anyone hoping to take advantage of Big Data. Unless the large amounts of data collected are made easily accessible, they'll remain limited in their use. Democratizing Big Data, that is, opening up access to it, is critical to gaining insight from it.

According to the Centers for Disease Control (CDC), heart disease is the leading cause of death in the United States, accounting for almost 600,000 deaths each year of the nearly 2.5 million in total.[20] Cancer accounts for just slightly fewer deaths. AIDS is the sixth leading cause of death among people

---

[17]http://obf.cancer.gov/financial/factbook.htm
[18]http://www.webmd.com/prostate-cancer/features/fifty-years-of-milestones-in-cancer-research
[19]http://www.cancer.gov/ncicancerbulletin/100212/page7
[20]http://www.cdc.gov/nchs/fastats/deaths.htm

aged 25 to 44 in the United States, down from the number one cause in 1995.[21] About two-thirds of all deaths in the United States each year are due to natural causes.

What about that much less serious, but far-reaching illness, the common cold? It's estimated that people in the United States catch a billion colds each year. That's three colds for every person. The common cold is caused by rhinoviruses, some 99% of which have been sequenced, and the number of different strains has historically been the reason the common cold is so hard to cure.

Although there's no cure on the immediate horizon, scientists have found commonalities in the proteins that make up the different forms of the virus, which may lead to advances in the future.

■ **Tip** The promise of Big Data is nowhere more prominent than in the area of medicine. Considering healthcare is about 20% of the GDP in the United States, smart product developers and entrepreneurs are looking for ways to optimize healthcare delivery, improve outcome success rates, and uncover important trends.

There are more than seven billion people living on the planet, according to estimates by the U.S. Census Bureau and the United Nations Population Fund.[22] Big Data applied to healthcare isn't just about addressing non-natural causes of death. It's also about increasing access to healthcare, improving quality of life, and reducing the costs associated with lost time and productivity due to poor health.

According to the most recent published CDC statistics, as of 2011, the United States spent about $2.7 trillion on healthcare annually or about $8,680 per person.[23] As people continue to live longer and fewer die young, more people are grappling with chronic illnesses and diseases that strike later in life.[24]

More children are receiving vaccines that are reducing death under the age of five, while outside of Africa, obesity has become a greater problem than malnutrition. In research that the Bill & Melinda Gates Foundation funded along with others, scientists found that people around the world are living longer, but they're also sicker. All of this points to the need for more efficiency in delivering healthcare and in helping people track and improve their own health as much as possible.

---

[21]http://www.ncbi.nlm.nih.gov/pubmedhealth/PMH0001620/
[22]http://en.wikipedia.org/wiki/World_population
[23]http://www.cdc.gov/nchs/fastats/health-expenditures.htm
[24]http://www.salon.com/2012/12/13/study_people_worldwide_living_longer_but_sicker/

Big Data and Where We Live: Energy, Leisure, and Smart Cities

Big Data isn't just improving health and well-being by changing the way we live, it's also changing the environments in which we live. Smart cities hold the promise of helping cities better organize for growth, according to the World Bank. The promise of smart cities[25] "is their ability to collect, analyze, and channel data in order to make better decisions at the municipal level through the greater use of technology." When it comes to "urban data, things haven't evolved much over the last century. We are today where we were in the 1930's on country or national level data," according to Christopher Williams of UN Habitat.

Today, more than half of the world's population lives in urban areas and that number is expected to rise to three quarters of the population by 2050. One challenge in collecting data from cities is standardizing the kind of data that is collected. Cities use a diverse set of data-collection methods and collect different kinds of data, making it difficult to compare data from multiple cities to develop best practices. But cities that gather relevant data can make better decisions about infrastructure investments. That's important given how long such investments tend to last.[26]

Devices like smart energy meters are already measuring energy consumption and providing consumers with detailed reports on their energy usage. In cities like San Francisco, smart parking meters report about the availability of parking on city streets, data which is then accessible to drivers via easy-to-use mobile apps. Those same parking meters work with products from PayByPhone to enable people to pay for parking by calling a phone number or using a mobile application. Some three million people now make use of the company's offering in 180 cities, including London, Miami, Vancouver, and San Francisco.

Meanwhile, applications are making it easier to get around major cities. CabSense analyzes data from the New York City Taxi & Limousine Commission and other sources to tell users the best corner on which to catch a cab based on the day of the week, the time of the day, and their location. CabSense analyzed tens of millions of historical data points and uses the data to rate every street corner on a scale of one to five.

Other apps tell users the best ways to make use of public transportation, and even the best train car to be in to make the fastest exit from the subway. Through a combination of applications that cities provide or at least sponsor the development of, and private application developers' innovative uses of

---

[25] http://web.worldbank.org/WBSITE/EXTERNAL/TOPICS/EXTSDNET/0,,conten tMDK:23146568~menuPK:64885113~pagePK:7278667~piPK:64911824~theSiteP K:5929282,00.html

[26] http://mercuryadvisorygroup.com/articles/sustain/100-pt4-reischl.html

publicly available data, cities are becoming easier to navigate and municipal governments are getting more information about what services will be most helpful to city residents. Even those cities without official smart city mandates or programs are getting smarter.

■ **Tip** Technology based on Big Data offers huge opportunities for those who can use it to make life easier or less expensive or more friendly to the global environment.

Big Data in Retail

Of course, where there are cities, there are consumers. Retailers are using Big Data not just to optimize inventory but to deliver personalized shopping experiences to their customers. Loyalty programs have been around for decades. But now, new Big Data Applications combined with in-store mobile technology like Apple's iBeacon can deliver advertisements, coupons, and deals to consumers who are in close proximity to particular products.

Retailers are using Big Data in other ways as well. Big Data enables retailers to pinpoint the best locations for new stores. Queue analytics enable retailers to evaluate shopping behavior within those stories, identifying bottlenecks and improving the shopping experience by reducing wait times. Shopper activity maps provide visualizations of consumer activity and the paths they take through stores, enabling stores to better optimize layouts and product displays to meet consumer needs.

■ **Tip** Amazon is rapidly eating the traditional retailer's lunch. The company is using Big Data to recommend products to customers, improve inventory management, and introduce new services. Traditional retailers need to get smart quickly about their use of Big Data, both to remain competitive and to keep loyal customers happy.

Retailers are using analytics to reach their customers outside the physical store as well. Big Data analytics can now deliver personalized deals and custom offers to shoppers to bring them back to the retailer's store or web site more frequently. Big Data isn't just changing how companies market to consumers—it's changing the nature of the conversation by delivering a shopping experience that's personalized to each consumer.

Big Data in Finance

Big cities mean big money and banks and other financial institutions are getting in on the Big Data wave as well. Banks use Big Data analytics to reduce fraud, leveraging pattern matching to detect purchases that seem out of the ordinary for a credit or debit cardholder. At the same time, financial institutions are using high-speed trading systems combined with complex algorithms to automate trading. These systems combine Big Data with the cloud to make automated buying and selling decisions involving billions of dollars every day.

Meanwhile, lenders are making more informed lending decisions based on wider sources of data. In addition to using traditional credit scoring approaches, lenders can now evaluate data sources like online reviews to figure out which restaurants are doing well and therefore might be good candidates to extend credit to—and which aren't. They can even combine that information with online real estate data to discover up-and-coming neighborhoods that may be good markets for new business. By combining multiple, previously unavailable data sets with the right algorithms, financial institutions can bring credit and banking to a broader set of consumers and businesses, supporting further growth.

The Mobile Future

Nearly five decades after the concept was introduced on *Star Trek* in 1966, the possibility of a handheld medical tricorder is becoming a reality.[27] Smartphone applications can now measure our heart rates and stress levels. Low-cost smartphone add-ons can gauge glucose levels and even provide ultrasounds at home. Such consumer applications and devices hold the promise of making some aspects of healthcare, at least health monitoring, more widely available and cost-effective. The data that such devices generate is useful to patients, to doctors performing diagnosis, and to scientists who rely on large quantities of data to inform their research.

In the enterprise software world, Software as a Service (SaaS) applications disrupted the traditional software delivery model by making applications like Customer Relationship Management (CRM) fast to set up and easy to use. Similarly, Big Data health applications that combine smartphones, low-cost hardware, and web-based analysis software have the opportunity to disrupt traditional, hard-to-use, and expensive medical devices, while improving the quality and reducing the cost of patient care at the same time.

---

[27]http://www.economist.com/news/technology-quarterly/21567208-medical-technology-hand-held-diagnostic-devices-seen-star-trek-are-inspiring

Tip In the enterprise, Software as a Service applications disrupted the traditional software delivery model by making applications fast to set up and easy to use. Big Data entrepreneurs have a similarly disruptive opportunity when it comes to the delivery model for healthcare.

Mobile devices may be one of the easiest ways for smart cities to collect critical data that enables them to improve services and make better decisions about infrastructure investments. Smartphones combined with low-cost medical add-ons may be one of the lowest cost and most efficient ways to expand access to health technology.

Some estimates put the number of smartphones in use worldwide at more than a billion, and the addition of the next billion devices could come as soon as 2015.[28] Mobile phone connectivity is on the rise in sub-Saharan Africa, reaching a penetration rate of some 70% as of 2014,[29] and smartphones, according to one writer, are not far behind.[30] Such devices come with built-in connectivity, making it easy for them to report data back and receive updates.

As in other areas of Big Data, it is at the intersection of the growing number of relatively inexpensive sensors for collecting data—such as iPhones and the specialized medical add-ons being built for them—and the innovative Big Data software applications where some of the most promising opportunities lie in improving our daily lives, the cities we live in, and healthcare globally. Combined with the digitization of medical records and more intelligent systems that can give doctors better information, Big Data promises to have a big impact on our health, both at home and in the doctor's office.

---

[28]http://thenextweb.com/mobile/2012/10/17/global-smartphone-users-pass-1-billion-for-the-first-time-report/
[29]http://allafrica.com/stories/201406101485.html
[30]http://techcrunch.com/2012/06/09/feature-phones-are-not-the-future/

Big Data Opportunities in Education

The Rise of Big Data Learning Analytics

Netflix can predict what movie you should watch next and Amazon can tell what book you'll want to buy. With Big Data learning analytics, new online education platforms can predict which learning modules students will respond better to and help get students back on track before they drop out.[1] That's important given that the United States has the highest college dropout rate of any OECD (Organisation for Economic Co-operation and Development) country, with just 46% of college entrants completing their degree programs.[2,3] In 2012, the United States ranked 17th in reading, 20th in science, and 27th in math in a study of 34 OECD countries.[4] The country's rankings have declined relative to previous years.

---

[1]http://www.ed.gov/edblogs/technology/files/2012/03/edm-la-brief.pdf
[2]The Organisation for Economic Co-operation and Development is an organization whose mission is to promote policies that will improve economic and social well-being and that counts 34 countries among its members.
[3]http://www.reuters.com/article/2012/03/27/us-attn-andrea-education-dropouts-idUSBRE82Q0Y120120327
[4]http://www.oecd.org/unitedstates/PISA-2012-results-US.pdf

Many students cite the high cost of education as the reason they drop out. At private for-profit schools, 78% of attendees fail to graduate after six years compared with a dropout rate of 45% for students in public colleges, according to a study by the Pew Research Center.[5]

Among 18 to 34 year olds without a college degree, 48% of those surveyed said they simply couldn't afford to go to college. Yet 86% of college graduates say that college was a good investment for them personally.

The data tells us that staying in school matters. But it also tells us that finishing school is hard. Paul Bambrick-Santoyo, Managing Director of Uncommon Schools, Newark and author of *Driven By Data: A Practical Guide to Improve Instruction*, has shown that taking a data-driven approach does make a difference.

During the eight years in which Bambrick-Santoyo has been involved with the Uncommon Schools, which consist of seven charter schools focused on helping students prepare for and graduate from college, the schools have seen significant gains in student achievement, reaching 90% proficiency levels on state assessments in many categories and grade levels.[6]

Using a data-driven approach can help us teach more effectively. At the same time, technology that leverages data can help students with day-to-day learning and staying in school. Netflix and Amazon present us with offerings we're more likely to buy, delivering a more personalized and targeted experience. Pandora figures out our music tastes and recommends new music to listen to. In the future, this kind of personalized experience won't just be used just for entertainment and shopping, but for education as well.

■ **Note**   Just as Pandora understands our musical tastes, tomorrow's education companies—built on Big Data analytics—will tailor custom educational experiences to specific students and their needs.

Adaptive Learning Systems

How can computers help students learn more effectively? Online learning systems can evaluate past student behavior, both for individuals and in aggregate, to predict future student behavior. Within a given course or courseware framework, an adaptive learning system can decide what material to show a student next or determine which areas a student might not yet fully understand. It can also show students, visually, how they are progressing with certain material, and how much material they've absorbed.

---

[5]http://www.pewsocialtrends.org/2011/05/15/is-college-worth-it/
[6]http://uncommonschools.org/bio/1017/paul-bambrick-santoyo

One of the strengths of an adaptive learning system is the built-in feedback loop it contains. Based on student interactions and performance, an adaptive learning system provides feedback to students, feedback to teachers, and feedback to the system itself, which humans or the system can then use to optimize the prediction algorithms used to help the students in the future. As a result, students, teachers, and educational software systems have a lot more visibility into what's going on.

Software can also predict which students are likely to need help in a given course.[7] Online courseware can evaluate factors such as login frequency and timeliness of turning in homework to predict whether students will pass or fail. Such software can then alert course instructors, who can reach out to students in danger of failing and offer them extra help or encouragement.

Knewton is one of the most well-known adaptive learning systems. Founded by a former executive of test prep company Kaplan, Knewton's system identifies strengths and weaknesses of individual students. The company started out by offering GMAT test prep, but now universities are using it to improve education.

Arizona State University (ASU), the country's largest public university with some 72,000 students, uses the Knewton system to improve students' proficiency with math. After using the system for two semesters with 2,000 students, ASU saw withdrawal rates drop by 56% and pass rates improve from 64% to 75%. The company has raised $105 million in venture capital and the World Economic Forum named Knewton a Technology Pioneer.

DreamBox, another provider of adaptive learning systems, is trying to improve math performance at the elementary school level. The company has raised $35.6 million in funding from well-known investors, including John Doerr of venture capital firm Kleiner Perkins, and Reed Hastings, CEO of Netflix.[8] DreamBox offers more than 1,300 lessons that help boost math proficiency. The company's applications are available for both desktop computers and Apple iPads.

At a broader level, data mining can also recommend courses to students and determine whether college students are off track in their selected major. ASU uses an eAdvisor system to coach students through college. The university's retention rate has risen from 77 to 84%, a change that the provost, Elizabeth Capaldi, attributes to eAdvisor.[9]

---

[7]http://www.nytimes.com/2012/07/22/education/edlife/colleges-awakening-to-the-opportunities-of-data-mining.html?_r=3&

[8]https://www.edsurge.com/n/2013-12-17-netflix-reed-hastings-leads-14-5m-series-a1-for-dreambox

[9]https://asunews.asu.edu/20111012_eAdvisor_expansion   and   http://www.nytimes.com/2012/07/22/education/edlife/colleges-awakening-to-the-opportunities-of-data-mining.html

The eAdvisor system tracks whether students fail key courses or don't sign up for them to begin with. It then flags such students so that academic advisors know to talk with them about their progress and recommend new majors, if necessary.

Such systems have vast amounts of data available to them, from individual course performance to standardized test scores to high school grades. They can compare data about any one student to data gathered about thousands of other students to make course suggestions.

This increased level of transparency extends all the way from students to teachers and administrators. Students get more information about their own progress. Teachers get more visibility into individual student progress as well as overall class progress, and administrators can look across all classes at a school to see what's working and what isn't. District administrators can then draw conclusions about what kinds of educational programs, software, and approaches are most useful and adjust curriculums accordingly.

Putting Education Online

One of the most interesting uses of Big Data as applied to education is the ability of adaptive learning systems to test many different educational approaches across a large number of students. Web sites use A/B testing to show one version of a web page to one visitor and another version to another visitor. Learning systems can do the same thing.

A learning system can evaluate whether students learn faster when they receive a lot of practice on a given type of problem all at once or when that practice is spread out over time. Learning systems can also determine how much material students retain after a given period of time and tie that back to the learning approaches used.

The intersection of Big Data and education doesn't stop at understanding how students learn. New startups are bringing educational materials online and opening them up to a much larger audience in the process.

■ **Note** While Big Data educational startup opportunities rest on a foundation of analytics and insight into learning models, the most effective succeed through smart marketing that brings in masses of students and sustains interest.

Khan Academy is an online destination containing thousands of educational videos. Salman Khan, the founder of the site, originally started recording the videos himself and the site now has more than 6,000 lectures available on a variety of topics, including history, medicine, finance, biology, computer science, and more.[10] The site's videos are stored on YouTube and in aggregate have received more than 440 million views.

The site's approach is simple yet effective. In addition to thousands of short videos, which highlight the material being taught rather than the person teaching it, the site uses hundreds of exercises to help teach concepts and evaluate the level of each student's comprehension.

Codecademy, a startup that has raised $12.5 million in funding, is focused on teaching people how to write software programs. Unlike Khan Academy, which relies heavily on videos, Codecademy focuses on interactive exercises. The site provides a series of courses grouped together in the form of tracks, such as the JavaScript track, the Web Fundamentals track, and the Ruby track, which enable users to learn different programming languages.

It's a long way from there to being able to build and make money from your own iPhone app, but the site provides a great way to get started. Imagine using courses from either site to train data scientists or educate business users on how to use analytics software.

Major universities are also putting their courses online. Harvard University and MIT teamed up to form edX, a not-for-profit enterprise that features learning designed specifically for study over the web. The site's motto is "The Future of Online Education: for anyone, anywhere, anytime." A number of major universities now participate in the program. Along with MIT and Harvard, Columbia University, Cornell University, the University of California at Berkeley, The University of Texas, Wellesley, Georgetown University, and many others, are also participants.

University faculty members teach the classes, which typically consist of short lecture videos accompanies by assignments, quizzes, and exams. In addition to enabling these universities to deliver course material electronically, edX provides a platform for learning about how students learn. EdX can analyze student behavior to determine which courses are most popular and which result in the greatest learning. EdX has said it wants to teach a billion students, and the *MIT Technology Review* called offerings like edX the most important educational technology in the last 200 years.[11]

---

[10]http://en.wikipedia.org/wiki/Khan_Academy
[11]http://www.technologyreview.com/news/506351/the-most-important-education-technology-in-200-years/

How Big Data Changes the Economics of Education

As the *Technology Review* points out, online learning isn't new. Some 700,000 students in the United States already use distance learning programs. What's different is the scale at which new offerings operate, the technology used to deliver those offerings, and the low-cost or free delivery models.

As in other areas of Big Data, what's changed is not that Big Data never existed before, but the scale and cost at which it can be accessed. The power of Big Data is its ability not just to gather and analyze more data, but to open access to that data to a much larger number of people and at a much lower cost. Free and low-cost education offerings such as those from edX are called massive open online courses, or MOOCs for short.

■ **Note** While MOOCs have received a lot of press coverage, the completion rates for courses are very low. As a result, the design of these courses needs to be revisited and I predict that a next generation of MOOCs—data-aware MOOCs—will emerge in the years ahead.

In 2002, only about 9.6% of college students were enrolled in at least one online course. By 2013, 33% of students were, according to studies by Babson Survey Research Group. That means some 7.1 million college students are taking at least one course online every year.[12]

Another offering, Coursera, was started by computer science professors at Stanford University. The company originally launched with Stanford, Princeton, the University of Michigan, and the University of Pennsylvania. Now it offers courses from more than 80 universities and organizations. It has received some 22.2 million enrollments for its 571 courses, with students from 190 different countries.[13] Notably, Data Analysis is one of the top courses on the site, highlighting just how much interest there is in Data analytics.

Meanwhile, Udemy, which has the tagline "the academy of you," brings together online instruction from a range of CEOs, best-selling authors, and Ivy League professors. The site takes a somewhat less academic approach to its offering and many of its courses are about practical business issues, such as raising venture capital. Unlike some of the other sites, Udemy allows course creators to provide their courses for free or charge for them.

---

[12]http://www.babson.edu/news-events/babson-news/pages/140115-babson-survey-study-of-online-learning.aspx
[13]https://www.coursera.org/about/community

Udacity, founded by Google vice president and part-time Stanford University professor Sebastian Thrun, has the goal of democratizing education. The company's initial courses have focused primarily on computer science related topics, but it is continuing to expand its offerings. To date, the company has raised $48 million in funding with the goal of expanding its course marketplace into lots of markets outside the United States.

As existing academic institutions search for ways to remain relevant in an online world, it is clear that the proliferation of such digital offerings will offer insight into the most effective ways to deliver educational content. While you may not yet be able to get a degree from Harvard, MIT, or Stanford over the web, getting access to their materials as well as to materials from anyone with something to teach is becoming a lot easier. At the same time, courses from leading academics on Big Data-related topics are now available to everyone—not just students at the universities at which they teach.

Of course, online courses can't provide the same kind of social or physical experience that classrooms or laboratories can provide. Courses in biology, chemistry, and medicine require hands-on environments. And just as social encouragement and validation is important when it comes to exercising and dieting, it may also be important when it comes to learning. The most promising educational systems of the future may be those that combine the best of the online and offline worlds.

Virtual environments may also provide a way to bring offline experiences online. The Virtual Medical Environments Laboratory of the Uniformed Services University adapts leading edge technology to provide medical training through simulation. Such simulation environments take advantage not only of software but also of hardware that simulates actual medical procedures. These environments can also simulate the noise or distractions that medical personnel may experience in the real world.

Using Big Data to Track Performance

The U.S. government, across federal, state, and local governments spends about $820 billion per year on education. That doesn't count all of the investment made at private institutions. But it does mean that administrators want visibility into how school systems are performing, and new systems are providing that visibility, according to Darrel M. West of the Brookings Institution.[14]

---

[14]http://www.brookings.edu/~/media/research/files/papers/2012/9/04%20
education%20technology%20west/04%20education%20technology%20west

DreamBox, the adaptive learning systems provider, also provides visibility for administrators. In addition to delivering adaptive learning tools, it has a dashboard capability that aggregates data for administrators to view. Administrators can track student progress and see the percentage of students who have achieved proficiency.

At a government level, the U.S. Department of Education has created a dashboard that summarizes public school performance for the entire country. The interactive dashboard is available on the web at dashboard.ed.gov/dashboard.aspx.

States use a variety of systems to report on educational progress. The state of Michigan provides a dashboard at www.michigan.gov/midashboard that indicates whether performance is improving, declining, or staying the same in areas such as third grade reading proficiency, college readiness, and academic proficiency between grades three and eight.

The state's third grade reading proficiency, for example, improved from 63.1% of students during the 2007–2008 school year to 70.0% of students during the 2013–2014 school year. According to the site, this measure is a strong indicator of future academic success.

Such systems improve accountability and provide more visibility into educational performance, according to West. Much of the information that goes into dashboards like these already exists, but web-based systems that have simple user interfaces and easy-to-view graphics are a big step forward in making such data accessible and actionable.

■ **Note**   A big part of the opportunity in education for entrepreneurs lies not just in accessing data that was difficult to access in the past, but also in the product developer's ability to interpret the data for end-users and to show it in visually powerful ways.

Data mining, data analytics, adaptive learning solutions, and web dashboards all present opportunities to improve education and increase access to it. But one of the biggest challenges, states West, is the focus on "education inputs, not outputs." Quite frequently, schools are measured on seat-time, faculty-student ratios, library size, and dollars spent rather than on results. "Educational institutions should be judged not just on what resources are available, but whether they do a good job delivering an effective education," says West.

With that in mind, it's clear that the approach taken by today's MOOCs needs to be revisited. MOOC completion rates are abysmally low, with just 5% of edX registrants earning a certificate of completion.[15] The typical long-form

---

[15]http://theinstitute.ieee.org/ieee-roundup/opinions/ieee-roundup/low-completion-rates-for-moocs

course content is similar to traditional university lectures, but without the social nature of a local community of fellow students to help everyone stay engaged. But what looks like failure in one context is often opportunity in another. In this case, the failure of the first generation of MOOCs presents an opportunity for entrepreneurs to create a next generation of MOOCs that are more engaging and result in higher completion rates.

In particular, data-aware MOOCs could automatically detect if students are becoming disengaged and present different content or interactive quiz modules to keep students engaged. Data-aware MOOCs could predict, based on the behavior of past students, which students are likely to drop out and alert advisors, who can then communicate with the students.

Data-aware MOOCs could even use data to determine which kinds and pieces of content, such as videos and interactive modules, are most effective. Based on this data, educational content creators could improve their content, while savvy application developers would create interactive MOOCs that use data to create learning interfaces that are as engaging, exciting, and social as today's most popular video games.

Education faces many of the same challenges when it comes to Big Data as other areas. Incompatible technology systems make it hard for schools to aggregate data within schools let alone compare data across different academic institutions. For example, some schools use separate systems for tracking academic performance and attendance. Transforming complex data sets about educational performance into key metrics is critical to making such data actionable.

How Big Data Deciphers What We Learn

As discussed in the Introduction, data not only makes computers smarter, it also makes human beings smarter. But the biggest question of all when it comes to education and Big Data may be the fundamental question about education itself: how do we learn?

Different people learn in different ways. Some students do better with visual learning while others do better with hands-on studies or when they write things down. Psychologists spent much of the last century constructing theories about how we learn, but they made little actual progress.

About ten years ago, scientists started taking a different approach. They used neuroscience and cognitive psychology to study how the brain learns.[16] What they discovered was that our ability to learn is shaped in large part not by what is taught but by the effectiveness of the learning process. A more efficient learning process can result in more effective learning.

---

[16]http://ideas.time.com/2011/10/12/the-science-of-how-we-learn/

Learning consultant Clive Shepherd captured some of the key insights from a talk by Dr. Itiel Dror of Southampton University on the science of learning.[17] While pop psychology has it that we only use five to ten percent of our brains, in reality, we use the entire capacity.

One of the keys to understanding how we learn is to recognize that our brains have limited resources for processing the huge amount of data we receive through our different senses. Our brains rely on all kinds of shortcuts to avoid getting swamped—what is known as cognitive overload.

As a result, to make learning more efficient, teachers can provide less information or take a very careful approach to how they communicate information. People have an easier time absorbing information if there's less noise that goes with it. But less noise also means less context.

One of the shortcuts the brain uses is to group things together. Teachers can make learning more efficient by grouping material so that students don't have to. Another approach to reducing cognitive overload is to remove every word or picture that isn't necessary to a particular learning goal. Challenging the brain helps with learning; researchers found that students learn more when they try to read a book for the first time than when they try to read the same book again.

Of course, all that still doesn't answer exactly *how* we learn. To cope with the vast amount of information it receives, the brain does a lot of filtering. The brain has evolved over many years, and one of the first things it needed to do was deal with basic survival.

Our ancestors were a lot more likely to survive if they could remember dangerous situations, such as stalking the wrong prey, and avoid them in the future. Such situations were often associated with moments of high emotion. As a result, it is easier for us to remember information that's associated with high emotion, whether it is positive or negative.[18]

Past experiences also influence how we retain information. Scientists believe that our brains store information in a sort of filing cabinet-like approach. This is one of the reasons it's easier to add more information to an existing area we know—an existing base of learning—than to learn something new from scratch.

---

[17]http://clive-shepherd.blogspot.com/2007/02/science-of-learning.html
[18]http://www.illumine.co.uk/blog/2011/05/how-the-brain-absorbs-information/

Using Big Data to Speed Language Acquisition and Math Skill

According to educational consultant Dr. David Sousa, citing Dr. Keith Devlin at Stanford University, mathematics is the study of patterns.[19] Sousa argues that all too often math is taught as simply a series of numbers and symbols, without any discussion of how it applies to daily life. Since meaning is one of the criteria the brain uses to identify whether information should be stored long term, math students may struggle to understand the subject without meaning to give it context.

Devlin highlights a number of cases where math applies to real life. Using probability to determine odds, calculating the amount of interest you pay when you buy a car, and applying exponential growth curves to understand population changes are three such examples.

But math may have as much to do with the language we use to represent our numbers as with how we learn it. As Malcolm Gladwell talks about in *Outliers*, referencing Stanislas Dehaene's book, *The Number Sense*, the English numbering system is highly irregular. Unlike English numbers, which use words like eleven, twelve, and thirteen, Chinese, Korean, and Japanese numbers use a more logical and consistent approach: ten-one for eleven, ten-two for twelve, and so on.

As a result, Asian children learn to count a lot faster. By four, Chinese children can count up to forty while American children of the same age can only count up to fifteen. They only learn to count to forty when they're a year older, putting them a year behind their Chinese counterparts. Gladwell cites another example: fractions. In Chinese three-fifths is literally, "out of five parts, take three," which makes such quantities much easier to work with. The language matches the concept.

The implications don't stop there. The brain has a working memory loop that can store about two seconds of information at a time. Chinese numbers can, in general, be pronounced in a shorter span of time than their English counterparts, which means that if you think about math in Chinese, you can remember more numbers at a time.[20]

More math education is highly correlated with higher earnings. In a study by the Public Policy Institute of California, authors Heather Rose and Julian R. Betts found that students who had completed calculus courses had higher earnings than those who had only completed advanced algebra.[21] They in turn had higher earnings than people who had completed only basic algebra.

[19]http://howthebrainlearns.wordpress.com/author/clarkbarnett/
[20]http://www.gladwell.com/outliers/outliers_excerpt3.html
[21]http://www.ppic.org/content/pubs/report/R_701JBR.pdf

Higher-level math education is also associated with higher college graduation rates. As the authors point out, correlation is not the same as causation, but they conclude that math education is highly associated with both earnings and college graduation rates.

If there's one person who knows more about learning math than just about anyone else, it's Arthur Benjamin, Professor of Mathematics at Harvey Mudd College. Benjamin is best known for his ability to perform mathemagics, in which he multiplies large numbers together in his head and produces the correct result.

As Benjamin shows, math doesn't need to be boring. It can be fun and entertaining. Proving the point, his TED talk on mathemagics has received more than four million views. Benjamin has also authored a book, *Secrets of Mental Math: The Mathemagician's Guide to Lighting Calculation and Amazing Math Tricks* as well as a DVD entitled *The Joy of Mathematics*. In his book, Benjamin shares a number of shortcuts to doing complex math in your head. When it comes to numbers, Big Data can be fun if given the right context!

So what about language? According to research by Dr. Patricia K. Kuhl at the Center for Mind, Brain, and Learning at the University of Washington, as infants, we store a lot of information about speech and language before we begin speaking. Simply listening to sounds helps our brains understand one language better than another.

Earlier, we talked about how the brain filters the vast amounts of information to which it is exposed. The infant brain does much the same thing with language. As infants master the language spoken by their caretakers, they ignore sound differences that aren't relevant.

For example, the different sounds for "r" and "l" are important in English (for words like "rake" and "lake"), but they aren't important in Japanese. Japanese babies tested at the age of six months could tell the difference between the two sounds equally as well as their American counterparts. By the age of 10 to 12 months, however, infants in the United States improved in their ability to tell the difference between the two sounds, while their Japanese counterparts got worse.

Kuhl attributes such changes to the infant brain focusing on the sounds it hears, the sounds of the infant's native language. During this period of rapid learning, it is also possible to reverse such declines by exposing infants to multiple languages. In one study, Kuhl had Chinese graduates students talk in Chinese with American infants. After 12 laboratory sessions, the American infants were able to recognize Chinese sounds just about as well as their Taiwanese counterparts. Kuhl concluded that the brains of infants encode and remember the patterns they hear well before those infants speak or even understand complete words.

By the age of six months, our infant brains are able to map the patterns of language having to do with vowels and consonants and by nine months, the patterns of words.[22] Kuhl describes the infant brain as analogous to a computer without a printer hooked up to it.

When it comes to reading, Kuhl's studies show that our ability to distinguish speech sounds at the age of six months correlates highly with language abilities like reading later in life. In other words, the better we are at distinguishing the basic building blocks of speech early in life, the better we are at complex language skills later in life.

According to Kuhl, we have about a trillion neurons (nerve cells) in place in our brains when we're born, but there are relatively few synaptic connections between. From the time we're born until about three years old, our brains form connections at a furious rate.

By age three, the brain of the average child has nearly twice as many connections as that of an adult. Moreover, the connections create three times more brain activity than in adults. At this point, the brain begins to prune unnecessary connections. Kuhl describes this as "quite literally like a rose bush, pruning some connections helps strengthen others." The pruning process continues until the end of puberty.

Are you out of luck if you don't start learning multiple languages at a young age? Common wisdom has it that learning a new language is difficult, if not impossible, after childhood. But one adventurous individual spent more than nine years traveling the world to see if how many new languages he could learn.

Much as mathematician Arthur Benjamin developed a set of shortcuts for doing rapid math calculations, Benny Lewis, author of the blog *Fluent in 3 Months: Unconventional Language Hacking Tips From Benny The Irish Polyglot*, developed a set of shortcuts for rapidly learning to speak new languages. Lewis, a former electrical engineering student with a self-proclaimed dislike of learning new languages, has shown that learning a new language as an adult is possible, if you take the right approach.

According to Maneesh Sethi, author of *Hack the System*, most of the challenge in learning a language later in life is that we go about learning it the wrong way. Sethi realized after studying Spanish for four years in high school that according to standardized tests he was an expert in Spanish. But, as he puts it, when it came to actually speaking Spanish, "I couldn't even order a burrito."[23]

---

[22]http://www2.ed.gov/teachers/how/early/earlylearnsummit02/kuhl.pdf
[23]http://lifehacker.com/5923910/how-i-learned-a-language-in-90-days

Sethi breaks the strategy of rapid language learning down into four steps: having the right resources, which include a grammar book, memorization software, and films/books; getting a private tutor; speaking and thinking only in the new language; and finding friends and language partners to converse with.

Sethi points out that by memorizing 30 words a day, you can learn 80% of the words necessary to communicate in a language in just 90 days. In Russian, for example, the 75 most common words make up 40% of occurrences. The 2,925 most common words make up 80% of occurrences, 75 less than the number of words you'll know by learning 30 new words per day. Sethi also highlights the importance of having the right mentality. Instead of thinking of himself as a blogger who wanted to learn Italian, he started thinking of himself as an "Italian learner (who blogs in his extra time)".

Fortunately, modern technology helps in many of the key areas, from memorization to tutoring to finding language partners. On the Mac, the Genius app uses a spaced repetition approach to flashcards that chooses questions intelligently based on your past performance. The more often you make a mistake, the more often the app will test you on a given word. Many online sites now provides live, interactive tutoring sessions via webcam.

Dealing with Data Overload: How Big Data Helps Digest and Filter Information

As children, we learn language by hearing it and speaking it, not by studying textbooks in a classroom. It should be no surprise therefore that the same approach that Kuhl highlights as being critical for children also works for adults: intensive, regular listening to and speaking the language we want to learn.

As Sethi points out, to learn a new language you must be an active learner. "Most people allow themselves to be taught to, but you have to take an active role in asking questions."

To cope with the vast amount of information it receives, the brain uses pattern matching and other shortcuts to make decisions. In this context, the approaches taken by Lewis, Benjamin, and Sethi make perfect sense. Rather than waiting for the brain to develop new pattern matching approaches and turn those into shortcuts, which is hard work, the key is to teach the brain new shortcuts instead.

"The challenge isn't in learning a new language, but rather learning how to learn a language," says Sethi. The same may hold true for other areas of learning.

> ■ **Tip** Product developers can strike gold when they uncover new ways of learning, rather than simply clothing old teaching methods—like rote memorization—in new apps. Each subject has unique content and data aspects you can discover, pursue, and ultimately package into a product that radically improves the learning experience.

Ultimately, learning is about taking in information and storing it, then drawing connections between what is already known and new information.[24] When it comes to Big Data, one takeaway from research on the brain and how we absorb information is that more data may give us more insight, but ultimately to be useful it needs to be digested and filtered down to a set of insights that are actionable; insights that can have a direct impact on our decisions.

As applied to education, Big Data is already helping to keep more students in school by figuring out when they're going to drop out. Adaptive learning solutions, whether in the form of complete systems or simple digital flashcard apps like Genius, are helping us learn more efficiently.

But there are many more possibilities. Much as our brains filter large amounts of information in order to make sense of it, Big Data technologies are also being designed to do the same. As more information is generated—and more of that information is in the form of unstructured data that is hard for traditional approaches like Master Data Management (MDM) to make sense of—new approaches are required.

Semantic search technology attempts to match searcher intent with the contextual meaning of vast amounts of data. Semantic search goes beyond simple word matching and tries to understand both the searcher's concepts and the concepts represented by documents, web pages, and even images and videos.

For example, if you type the word "flights" into the Google search box, Google uses semantic intent to predict what you're actually searching for.[25] In the case of the word "flights," Google might combine that search term with knowledge of your current location to suggest flights originating in your city. Such semantic searching techniques can help us tackle the information overload problem that is growing bigger every day.

Related to semantic search, semantic filtering is the application of intent and conceptual understanding algorithms to the filtering of vast quantities of information, such as news sources. Rather than filtering and suggesting relevant content to read or view based on simple word match techniques, semantic filtering analyzes the concepts contained in news articles, combined with

---

[24]http://www.stanford.edu/class/ed269/hplintrochapter.pdf
[25]http://searchengineland.com/5-ways-to-unlock-the-benefits-of-semantic-search-hummingbird-175634

knowledge of a user's preferences or interests, to recommend highly relevant content. The underlying algorithms can become smarter over time about which content to recommend based on feedback from users about which content they want to see more of—and which they don't.

We are already integrating such techniques into our daily routines through the use of recommendation systems, spam filtering, news filters, and search boxes. There are opportunities to build many more applications that help us filter the data overload into actionable, relevant information.

At the same time, Big Data can also help us with the actual practice of education. To be truly successful, data-driven educational applications will need to take lessons from today's video games and social networking applications. Facebook is able to drive some 40 minutes of voluntary engagement per day while games like World of Warcraft engage players for hours at a time.

Both of these applications have cracked the code on delivering content that engages—and optimizing the display of that content over time—as well as providing a social experience driven by status and recognition. By combining the right aspects of social, content, and Big Data, application developers can apply the same principles to produce the break-through educational appellations of tomorrow. From enabling more students to graduate to making it possible for us learn more efficiently, Big Data holds the promise of helping us become not only better teachers but better students as well.

Capstone Case Study: Big Data Meets Romance

Products, Services, and Love: How Big Data Plays Out in the Real World

When it comes to developing new applications to help people meet, connect, and interact with each other, the opportunities are limitless. You can combine interesting new sources of data, such as people's interests in music, movies, travel, or restaurants, with new technologies like mobile applications. For example, you could build an application that aggregates and analyzes data about people's interests and then alerts them when they are near each other so they can meet up in person. Or you could build an application that monitors people's activity levels, such as their sleep schedules, how active they are, and how many meetings they are in and warn them if their stress level is increasing—which could have a negative impact on their personal relationships.

Calendar-based applications could use data to figure out if you've been neglecting to spend time with your friends, while new social networking applications could make it easier to connect with people you care about while on the go.

A number of companies are taking advantage of data and mobile technologies to help us be more connected and to keep users more engaged. Social networking site Facebook uses Big Data to figure out what content to show, which ads to display, and who you should become online friends with. Dating site Match.com stores some 70 terabytes of data about its users.[1] Recent mobile dating entrants Tinder and Hinge use Big Data to connect people based on their Facebook profiles.[2]

Online dating web site OkCupid, which was acquired by Match.com in 2011, conducted a series of research studies on what makes for the most successful online dating profiles. In a post entitled "The 4 Big Myths of Profile Pictures," the OkCupid researchers described the results of analyzing more than 7,000 profile photos to figure out which ones produced the best results, as measured by the number of messages each user received.[3]

Of course, just receiving more messages isn't necessarily a good thing. Quality is often more important than quantity. But before we explore that issue, let's first take a look at what the researchers found.

The researchers characterized photos into one of three categories: flirty, smiling, and not smiling.

The researchers discovered that for women, eye contact in online profile pictures was critical. Those with a flirty look got slightly more messages each month than those without and significantly more messages than those who were characterized as not smiling.

Without eye contact, the results were significantly worse. Regardless of facial expression, those who didn't make eye contact received fewer messages overall than those who did.

In contrast to women, men had the most success in meeting women when they used profile pictures in which they looked away from the camera and didn't smile. However, men who used photos that showed them looking away from the camera and flirting had the least success in meeting women. For both men and women, then, looking away from the camera and flirting produces the worst results.

The researchers also found that for men, wearing "normal clothes" in a photo or being "all dressed up" doesn't make much of a difference when it came to meeting a potential mate.

In perhaps one of their most interesting findings, the researchers discovered that whether or not someone's face appears in a photo doesn't have a big impact on how many messages they receive.

---

[1]http://www.bbc.com/news/business-26613909
[2]http://www.cnbc.com/id/101151899
[3]http://blog.okcupid.com/index.php/the-4-big-myths-of-profile-pictures/

Photos of people in scuba gear, walking across the dessert, or with their faces out of the camera completely were about as likely to help users of the site generate interest as those photos in which users showed their faces. Women who used photos showing their faces received, on average, 8.69 contacts per month, while those who didn't show their faces received 8.66 messages per month. Men who used facial photos met 5.91 women for every 10 attempts, while those not showing their faces met 5.92 women for every 10 attempts. This remarkable data, as shown in Figure 12-1, dispels the myth that you should make sure your face is showing in online dating profile photos.

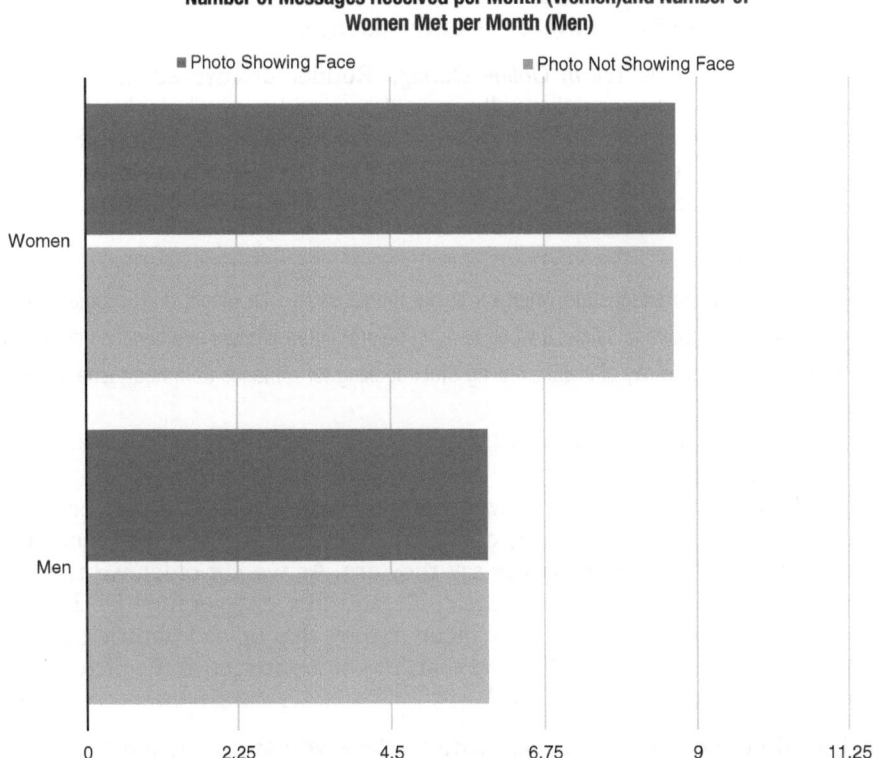

Figure 12-1. Dispelling the myth that you should make sure your face is showing in online dating photos

The researchers' conclusion? That the photos do all the work—they "pique the viewer's curiosity and say a lot about who the subject is (or wants to be)." It's important to note as the authors of the study stated, "we wouldn't recommend that you meet someone in person without first seeing a full photo of them."

What all that means is that Big Data isn't just a tool for business. Given the right data sources it can also give us insights into how best to portray ourselves to find the right mate. The other takeaway, of course, is that you shouldn't underestimate the importance of choosing the right photo—based on the data.

How Big Data Outs Lies, Damn Lies, and Statistics

Of course, people are known to stretch the truth when it comes to online dating. In another study, OkCupid co-founder Christian Rudder looked at data from some 1.51 million active users of the dating site.

In *The Big Lies People Tell In Online Dating*,[4] Rudder discovered that when it comes to height, "almost universally, guys like to add a couple inches." In fact, as guys "get closer to six feet," they round up more than usual, "stretching for that coveted psychological benchmark." Women also exaggerate their height, "though without the lurch toward a benchmark height."

Note   People often exaggerate when on social media sites, with height and income among the qualities users choose to inflate. It won't be long before astute product and service developers find ways to test the veracity of claims, saving those looking for romance from heartache, at least occasionally.

What about income? Do people who claim to make $100,000 a year or more really make that much? When it comes to online dating, Rudder found that people are 20% poorer than they say they are. As we get older, we tend to exaggerate more, with those in their 40s and 50s exaggerating by 30% or more. People don't just alter their income; they also upload photos that are out of date. Rudder's research found that "the more attractive the picture, the more likely it is to be out of date."

So how do the dating sites decide who to show you as potential matches?

Match.com, which has some 1.8 million paying subscribers, introduced a set of new algorithms codenamed Synapse to analyze "a variety of factors to suggest possible mates," according to an *FT.com* article, "Inside Match.com."[5]

What we say we're looking for isn't always what we're actually looking for. Although the algorithm takes into account people's stated preferences, it also

---

[4]http://blog.okcupid.com/index.php/the-biggest-lies-in-online-dating/
[5]http://www.ft.com/intl/cms/s/2/f31cae04-b8ca-11e0-8206-00144fcabdc0.html

factors in the kind of profiles that users actually look at. For example, if a user states a preference for a certain age range but looks at potential mates outside that range, the algorithm takes that into account and includes people outside the range in future search results.

What makes the challenge of predicting preferences even more complicated is that unlike with movie or book recommendations, algorithms that match human beings together have to take into account mutual preferences. "Even if you like *The Godfather, The Godfather* doesn't have to like you back,'" Amarnath Thombre, head of analytics at Match.com was quoted as saying.

The challenge with such algorithms is that although Match.com has data on the 75 million users who have registered on the site since its founding in 1995, it doesn't have much data on which dates are successful and which ones aren't. This inability to close the loop is an important missing element in creating the ultimate matchmaking algorithm.

That's why when people cancel their subscriptions, dating sites often ask whether the reason is because they're dissatisfied with their online dating experience or because they met someone. Not only is such data useful for marketing, it can also factor into predictive algorithms.

Using Big Data to Find the Missing Data

But it is just that kind of missing data that may be causing machine-matching algorithms to fail. As with all Big Data analytics and predictive engines, such algorithms are only as good as the data that engineers used to develop them and the data they're fed. If key data is missing, predictive algorithms won't work.

Eli J. Finkel from Northwestern University and Benjamin R. Karney at the University of California, Los Angeles, the authors of a study published in the journal *Psychological Science in the Public Interest*, point out in their study and in a corresponding *New York Times* opinion piece that what really matters is how people interact when they meet each other in person, not what they say online.[6]

"Things like communication patterns, problem-solving tendencies, and sexual compatibility are crucial for predicting the success or failure of relationships," say the two professors.

They also point out that the way in which "couples discuss and attempt to resolve disagreements predicts their future satisfaction and whether or not the relationship is likely to dissolve." These, however, are not the sorts of

---

[6]http://www.nytimes.com/2012/02/12/opinion/sunday/online-dating-sites-dont-match-hype.html

characteristics that are easily evaluated in the context of an online dating site. Moreover, dating sites don't take into account the environment of a relationship, including stresses such as "job loss, financial strain, infertility, and illness."

The authors also point out that while dating sites may collect a lot of information, such information is a very small piece of the pie when it comes to figuring out what will make two people a good long-term match. While many sites claim to match people based on common interests, a 2008 study of 313 other studies found that "similarity on personality traits and attitude had no effect on relationship well-being in established relationships."

A 2010 study of more than 23,000 married couples showed that having major personality attributes in common, such as neuroticism, impulsivity, and extroversion, only accounted for half a percent of marriage satisfaction, meaning that 99.5% was due to other factors.

The conclusion of the studies' authors? Online dating isn't any better or worse than any other way to meet potential mates. While the algorithms online dating sites use may make for good marketing, such algorithms may ultimately just be a way to help users of such sites get started with their online dating experience, and to provide a manageable pool of potential mates in densely populated areas such as New York City.

What such research really points out from a Big Data perspective is the importance of having complete data. Whether algorithms are trying to recommend movies, tell sales people which customers to call next, or suggest potential mates, the algorithms are only as good as the data. Without sufficient data and a way to close the loop by knowing whether an algorithm made a correct prediction, it's difficult to create accurate algorithms. Like throwing darts at a dartboard, getting all the darts in the same area doesn't matter if that area isn't the bull's-eye. So capturing complete data—and being able to close the loop about what works and what doesn't, is key.

Predicting Marriage Success with Big Data

When it comes to prediction, one scientist is able to predict with astounding accuracy which matches are likely to succeed. Professor of psychology Dr. John Gottman is famous for running a physiology lab known as the "Love Lab" at the University of Washington.

Gottman has studied more than 650 couples over a period of 14 years. Based on his research, Gottman can predict with over 90% accuracy based on a half hour interview with a recently married couple whether their marriage will last.[7]

---

[7]http://www.uwtv.org/video/player.aspx?pid=OJG2Xv9kB9nO

Gottman refers to what he calls mental map-making as the basis for romance. Mental map-making, in the context of a relationship, is the process of finding out about our partners. One simple example of mental map-making Gottman provides is of men who have an interest in what their wives are going to do on a given day.

An active mental map-maker not only gathers information but also thinks about that information during the day and follows up on it later. That means asking about a spouse's meeting, lunch, or an event they talked about in the morning.

Remarkably, the process of gathering data, thinking about that data, and following up on that data is useful not only in computing, but in maintaining a healthy relationship as well.

Gottman and his colleague, Professor James Murray, took things even further and developed Big Data models of biological research. They created mathematical models of human behavior that they used to analyze and predict marriage success. The two professors, along with several others, even published a book entitled *The Mathematics of Marriage: Dynamic Nonlinear Models*.

Gottman and his fellow authors believe that "the development of marriage is governed, or at least can be described, by a differential equation."[8] Gottman and his colleagues are able to describe a marriage using the mathematics of calculus and simulate how couples will act under various conditions.

Gottman's research discovered four negative behaviors that frequently predict divorce: criticism of a partner's personality, contempt, defensiveness, and emotional withdrawal from interaction.[9] Gottman describes contempt, in particular, as "sulfuric acid for love."

By combining an evaluation of a couple's responses to various questions with analysis of body language and biological data via mathematical models, Gottman can not only predict whether a marriage is on course to succeed, but can also suggest immediate changes for course correction if it isn't.

Given all this talk of mathematics and differential equations, it might seem like Gottman is taking the magic out of romance, but as Gottman puts it, it's important for scientists and researchers to have an objective model for measuring human relationships. And the information gleaned from such Big Data models can help keep relationships strong.

---

[8] http://www.slate.com/articles/life/do_the_math/2003/04/love_by_the_numbers.2.html
[9] http://en.wikipedia.org/wiki/John_Gottman and http://psycnet.apa.org/?&fa=main.doiLanding&doi=10.1037/0893-3200.14.1.42

Gottman studied those who remained happy with their marriages to figure out what kept their relationships happy. His conclusions were that couples that focused on preserving knowledge of each other, mutual admiration, and affection for each other were more likely to be happy than those who did not. In particular, he found that couples had to be five times as positive with each other as negative.

Can failing marriages be saved? According to Gottman's analytical research, they can. Many couples that go through something called marriage counseling relapse, falling into the same old habits, after two years.

However, based on his studies, Gottman concluded that there are two active ingredients that can produce a lasting positive effect on a marriage. First, reduce negativity in conflict resolution. Second, increase overall positivity by focusing on helping partners in a marriage "be better friends." Most importantly, Gottman's research isn't just a lot of data points. As with all of the best Big Data analyses, Gottman has converted his research into simple, yet powerful actionable insights. In this case those insights can lead us to happier, more fulfilling relationships.

■ **Note**   The best Big Data analyses, and those that can be turned into successful products most easily, are those that convert heavy-duty research into simple, powerful, actionable insights.

How Facebook Is Shaping Relationships

While Gottman and his colleagues are helping couples in the offline world based on their years of research, perhaps no online company is having a bigger impact on relationships than social networking giant Facebook.

The company, which now has more than a billion users worldwide, is the go-to destination for photo sharing, status updates, and a timeline of your life, both on your own and with others.

Social networking sites like Facebook represent such relationships in something called the social graph. Unlike in a one-to-one relationship, a social graph consists of many interconnected relationships. If Joe knows Fred and Fred knows Sarah, then in the context of the social graph, Joe has a link to Sarah via Fred. But such interconnections extend even further with the inclusion of interests, places, companies, brands, birthdays, and status updates, among other non-human elements. In the context of the social graph, people have connections not just with other people, but with activities, events, companies, and products.

Tip Just as some companies are focused on building databases for structured and unstructured data, there are also companies that build databases specifically for graph-based data. For example, Neo4j has focused on building a database specifically to represent connections. The company has raised some $24 million to date and counts companies like eBay, HP, and Walmart among its customers.

The concept of six degrees of separation has popularized the notion that two people are at most six steps away from each other in terms of social connections. Business networking site LinkedIn takes advantage of this concept by showing business people how they're connected to other people through intermediary relationships. These professionals then take advantage of this connectivity when they want to talk with people they don't know directly but who are a friend of a friend or a colleague of a colleague.

At Facebook, people are even closer to each other than the six degrees concept might suggest. Sanjeev Kumar, an engineering manager at the company, indicated that Facebook users are highly connected, with an average 4.74 degree of separation between them.[10] Relationships in the social graph are made denser due to connections with places, interests, and other elements.

By representing the relationships between a very large number of connections, the social graph can answer lots of interesting Big Data questions. That includes not only questions posed by data analysts, but questions actively or passively posed by hundreds of millions of users, such as who should I connect with?, what photos should I look at?, and which information is important to me?

Although most of these users don't think of themselves as performing queries against the Big Data that is the social graph, that is exactly what they're doing, or at least what social networking services are doing on their behalf.

Social graphs are equally interesting in Big Data from a technical perspective. Answering questions such as the ones above consumes a lot of computing resources. Each query involves working with a very large subset of the overall graph (known in technical terms as the working set) and is highly customized to each user.

Tip Social graphs are the digital representation of our human connections. They may be one of the most powerful sources of insight into human relationships.

---

[10]http://www.ece.lsu.edu/hpca-18/files/HPCA2012_Facebook_Keynote.pdf

What's more, the social graph represents lots of actual data, not just the interconnections in the graph itself, but the photos, videos, status updates, birthdays, and other information associated with each user. A query has to return the right set of relationships. It also has to return the data associated with those relationships and do so near instantaneously.

The social graph represents so much data that, to keep up with it, Facebook has had to develop custom servers, build out its own data centers, and design special software for querying the graph and storing and retrieving its associated data efficiently.

So what does all this mean for our relationships?

Of the many ways to express ourselves on Facebook, "relationship status is the only one that directly involves another person."[11] We commonly announce engagements, marriages, breakups, and divorces on the social networking site.

In 2010, about 60% of Facebook users set a relationship status on their profile and as of December of that year, women outnumbered men by a rate of 1.28 to 1.[12] As of January, 2014, about 53.3% of the site's users were female while 45.6% were male.[13] In 2011, a third of all divorce filings mentioned the word Facebook, up from 20% the previous year.[14]

Announcing our relationship status online can strengthen what researchers refer to as "feeling rule reminders."[15] Such rules are the social norms that tell us when and what to feel, and how strong our emotions should be. By declaring our relationship status online we reinforce such rules.

Social networking sites like Facebook can also affect one's health and personality. Some people present themselves authentically on the site, while others present themselves in an improved light due to feelings of insecurity.[16]

About 24% of Americans and 28% of Brits have admitted to lying or exaggerating on a social network about what they've done and/or who they've met, according to statistics cited by Cara Pring of social networking statistics site

---

[11]http://www.time.com/time/business/article/0,8599,1895694,00.html
[12]http://blogs.channel4.com/benjamin-cohen-on-technology/the-way-facebook-changes-relationships/2541
[13]http://istrategylabs.com/2014/01/3-million-teens-leave-facebook-in-3-years-the-2014-facebook-demographic-report/
[14]http://abcnews.go.com/Technology/facebook-relationship-status/story?id=16406245#.UMzUJaWTVjY
[15]http://doctorat.sas.unibuc.ro/wp-content/uploads/2010/11/Issue2_GregBowe_Romantic.pdf
[16]http://allfacebook.com/infographic-facebook-hurt-relationships_b45018

The Social Skinny.[17] Meanwhile, visiting our profiles too frequently can make us overly aware of ourselves, causing stress and anxiety.

Having more digital friends than others—the average Facebook user has 229 friends on the social network—may give us a shot of self-confidence due to receiving additional social support. Some 25% of people believe that social networks have boosted their confidence. Having fewer friends than others may make us feel self-conscious, but the smaller digital circle may also lead to more genuine interactions.

In case there is any doubt about how big an impact online socializing is having on our lives, 40% of people spend more time socializing online than they do having face-to-face interactions. Internet users spend 22.5% of their online time engaged in social networking activities, with more than half of all Facebook users returning to the site on a daily basis.

■ **Note** If you're worried you may not have enough of a market for your social media application, bear in mind that nearly half of the people who socialize online spend more time there than interacting with human beings in the real world.

How Big Data Increases—And Decreases—Social Capital

We value our online identities and the relationships they entail not just because they are a way to express ourselves but also because we associate them with social capital. Social capital is the benefit we receive from our position in a social network and the associated connections and resources that are a part of that network, both online and offline.

Such capital comes in two forms, bonding and bridging, according to researchers at the Human-Computer Interaction Institute at Carnegie Mellon University and at Facebook.[18] Bonding refers to social capital derived from relationships with family members and close friends, while bridging refers to that derived from acquaintances.

Both have value, but bonding typically enables emotional support and companionship. Bridging, which comes from a looser and more diverse set of

---

[17]http://thesocialskinny.com/216-social-media-and-internet-statistics-september-2012/
[18]*Social Capital on Facebook: Differentiating Uses and Users* (May, 2011)

relationships, typically enables access to new communication and opportunities such as job openings, because very close ties are likely to share redundant information.

The researchers classified activities on social networks into three categories. The first kind is directed communication with individual friends via chat or wall posts. The second involves passive consumption of social news by reading updates from others. Finally there is broadcasting, which consists of updates not directed at any particular person.

The researchers concluded that directed communication has the potential to improve both bonding and bridging capital due to the rich nature of the content and the strength of the relationship between the two communicators. In the case of one-to-one communication, both offering and receiving information increases the strength of relationships. Moreover, simply communicating one to one, due to the effort required relative to broadcast communication, signals the importance of a relationship.

For bridging social capital, only one-to-one communication increases social capital for senders on Facebook. Other forms of communication increase social capital only for the receiver. As the researchers point out, undirected broadcasts and passive consumption of news and updates may increase knowledge for the recipients and consumers of the news, but they don't further the development of relationships.

Broadcast communication is useful as a source of information about others that we can then use to increase the strength of relationships or to develop new friendships by referencing shared interests. In terms of casual acquaintances, putting out broadcasts doesn't buy us much, but receiving broadcasts and consuming news may. (Ironically, of course, recipients can only derive such value by nature of senders making the broadcasts.) The benefit of receiving broadcast information is greater for those with lower existing social communication skills.

The researchers also found that only a narrow set of major life changes have a significant impact on bridging social capital. Moving, for example, has a positive impact on bridging social capital, likely due to adding new relationships that diversify our access to information and resources. Losing one's job, however, has a negative impact due to losing the social context associated with former coworkers.

Where does this leave us when it comes to the impact of changes in our personal relationships on our broader networks? Much as we worry about the impact of life changes such as marriage, divorce, death, family additions, new jobs, and illness on our broader social networks, the reality is that while such events have a big impact on us personally, they have relatively little impact on our broader social networks.

How does this compare to increases and decreases in social capital via Facebook? Every time we double our one-to-one communication in the network, that results in about the same quantity of change impact on bridging social capital as moving to a new city. It's equivalent to about half the impact associated with losing a job. In other words, a one-to-one digital message that establishes a new connection can have as much impact on bridging relationships as moving to a new city.

The implication is that, if we choose to take advantage of one-to-one communication online, social networks can greatly reduce the friction normally associated with expanding our broader networks. This may be one reason why when it comes to job searches, one-to-one messages on professional business networking site LinkedIn, even messages that come by way of a connection, are so effective.

While communication clearly plays a huge role in the development of relationships, so does information. Although its role may be somewhat less obvious than that of Facebook, Google impacts our relationships as well.

Ask nearly anyone who has been on a date that started online and they will tell you that they've Googled their potential mate. According to *It's Just Lunch*, a dating service, 43% of singles have Googled their date before going out with them. And of the 1,167 singles surveyed, 88% said they wouldn't be offended if their date Googled them.[19]

Type virtually anyone's name into the search engine and you'll see a series of results related to that person. They range from web sites providing background information to articles someone has been mentioned in or written, to work profiles on LinkedIn.

The relative newness of Facebook, Twitter, and other forms of digital social communication mean that academic research on the impact of such mediums is relatively limited. What is clear, however, is that one's online identity is a massive form of Big Data.

When it comes to humans, Big Data is the information we share about ourselves in the form of photos, videos, status updates, tweets, and posts, and about our relationships, not to mention the digital trail we leave behind in the form of web site clicks and online purchases.

Big Data and Romance: What the Future Holds

Enhancements to digital social communication, such as virtual gifts, may seem an unlikely way for people to communicate romantic interest, but such forms of expression are becoming more and more commonplace. Virtual items like

---

[19]http://www.itsjustlunch.com

flowers that look like actual real-world goods but exist only in digital form have become quite popular both in social games and on social networks. People have shown a willingness to pay for such digital items.

One can imagine a future in which recommendation engines can accurately predict which gifts will be most positively received and make suggestions to the senders. Of course, social networks might want to start by ensuring that partners simply don't forget each other's anniversaries!

Whether we want such insight or not, Big Data is both collecting and providing more information about us and our relationships. Algorithms such as those developed by Match.com to try to recommend better matches may be missing important data about how relationships evolve over time, but we can foresee a time when services integrated with Facebook and other platforms provide high-quality recommendations and predictions.

Indeed, Facebook may have more insight into our relationships over time than just about any other web site. Blog site *All Facebook* cites a passage from *The Facebook Effect: The Inside Story of the Company That Is Connecting the World*, saying, "By examining friend relationships and communications patterns [Zuckerberg, Facebook's co-founder and CEO] could determine with about 33% accuracy who a user was going to be in a relationship with" a week in advance. Zuckerberg was apparently able to use data about who was newly single, who was looking at which profiles, and who was friends with whom to glean such insights.[20]

Online dating and social communication are becoming more and more socially acceptable and are producing vast amounts of data in the process. Mass forms of communication are no substitute for one-to-one communication. But social networks that make such communication easier have the ability to reduce dramatically the amount of effort required to create new relationships and that required to maintain existing ones. Today algorithms such as those Match.com has developed may simply be a way to start communicating. But in the future, the opportunity for Big Data to improve our relationships, both personal and professional, is large.

As more people use smartphones like iPhone and Android devices, we're seeing mobile applications spring up for dating and as a more general way to meet new people. Applications such as Skout help users discover new friends wherever they are, whether at a local bar, at a sporting event, or while touring a new city. The application provides integrated ways to chat, exchange photos, and send gifts. Mobile applications will continue to reduce communication barriers and help us stay in touch with people we care about, all the while

---

[20]http://allfacebook.com/facebook-knows-that-your-relationship-will-end-in-a-week_b14374

generating immense amounts of data that can help further refine interactions and introduce new connections.

Big Data may not yet be able to help an ailing relationship, but it can give us insight into the context surrounding our relationships, such as when the most difficult times of year are for relationships, based on the number of breakups that happen at those times. By learning from such data we can take proactive steps to improve our relationships.

Big Data may also help us determine if a friend or relative is headed for trouble, by determining if he or she is communicating less frequently than normal, has an increased level of stress, or has experienced multiple major life events such as divorce, job loss, or the death of a relative or close friend that could lead to depression or other issues. At the same time, by continuing to make communication easier, Big Data may very well strengthen existing relationships and support the creation of new ones.

Our desire to create new bonds and strengthen existing ones—our inherently social nature as human beings—is one of the biggest drivers of the creation of new technologies to help us communicate more easily.

■ **Tip** The market for match-making apps is big. But the market for helping existing relationships thrive using Big Data may be even bigger.

We may not think of web sites like Facebook or Google as Big Data Applications because they are wrapped in easy-to-use, consumer-friendly interfaces. Yet in reality, they represent the enormous potential of Big Data.

Applications like Facebook are just the tip of the iceberg when it comes to applying Big Data to improve our relationships and form new friendships. By combining data with mobile, a variety of new applications are possible. Why stop at just matching people up based on general interests? Why not go a step further and recommend matches based on the music we've listened to recently, the movies we like to watch, or the food we like to eat? When it comes to building new relationship applications, today's entrepreneurs have the opportunity to be as disruptive to existing matchmaking and social networking services as the iPod was to the Walkman.

In all the discussion of the size and economics of Big Data, of the talk of real-time data streams and the lower costs of storing and analyzing data, it is easy to lose sight of the positive impact that Big Data has on our daily lives, whether we're talking about online matchmaking or taking steps to improve our relationships. Not only can Big Data open new avenues for you to find the love of your life, the good news for singles the world over is that in the future Big Data is also likely to help them hold onto that love once they find it.

Big Data Resources

General Information

- www.data.gov. The home of the U.S. Government's open data.

- Feinleib, David. "Actionable Insights from Big Data." www.youtube.com/watch?v=GoKTISLYs_U.

- McCandless, David. "Information Is Beautiful: Ideas, Issues, Knowledge, Data—Visualized!" www.informationis-beautiful.net/.

- McCandless, David. "The Beauty of Data Visualization." www.ted.com/talks/david_mccandless_the_beauty_of_data_visualization.

- Mayer-Schönberger, Viktor and Kenneth Cukier. *Big Data: A Revolution That Will Transform How We Live, Work, and Think.* John Murray Publishers, London: 2013.

- Rosling, Hans. "The Best Stats You've Ever Seen." www.ted.com/talks/hans_rosling_shows_the_best_stats_you_ve_ever_seen.

- Silver, Nate. *The Signal and the Noise: Why So Many Predictions Fail—But Some Don't.* Penguin Press, New York: 2012.

- Tableau Software data visualization gallery. www.tableau-software.com/public/gallery.

Big Data Software and Services

The Big Data Landscape provides the most comprehensive list of Big Data software and services. The following are great launching points for getting started with Big Data:

- Amazon Elastic MapReduce (EMR): http://aws.amazon.com/elasticmapreduce/

- Amazon Kinesis: http://aws.amazon.com/kinesis/

- Apache Hadoop: hadoop.apache.org/

- CartoDB: www.cartodb.com

- Cassandra: cassandra.apache.org/

- Cloudera: www.cloudera.com

- Google Cloud Dataflow: http://googlecloudplatform.blogspot.com/2014/06/sneak-peek-google-cloud-dataflow-a-cloud-native-data-processing-service.html

- Hortonworks: www.hortonworks.com

- MapR: www.mapr.com

- MongoDB: www.mongodb.com

- New Relic: www.newrelic.com

- QlikTech: www.qliktech.com

- Splunk: www.splunk.com

- Tableau Software: www.tableausoftware.com

Big Data Glossary

The following glossary defines some of the key terms used in the Big Data world. There are many more—and more to come as Big Data continues to evolve.

Amazon Kinesis—A cloud-based service for real-time processing of streaming Big Data such as financial data and tweets.

Apache Hadoop—An open source framework for processing large quantities of data, traditionally batch-oriented in nature.[1]

---

[1] http://hadoop.apache.org/

Apache Hive—Software for querying large datasets contained in distributed storage like Hadoop. Hive enables data querying using a SQL-like query language called HiveQL.

Batch—An approach to analyzing data in which data is processed in large chunks called batches.

Data analyst—A person responsible for analyzing, processing, and visualizing data.

Google BigQuery—A cloud-based service provided by Google for analyzing large quantities of data on Google's infrastructure, using SQL-like queries.

Google Cloud Dataflow—A cloud-based service for data integration, preparation, real-time stream processing, and multi-step data processing pipelines. Google positions Cloud Dataflow as the successor to Hadoop and MapReduce.

HDFS—Hadoop distributed file system, a distributed file system for storing large quantities of data that run on commodity hardware.

Machine data—Data such as system logs generated by machines like computers, network equipment, cars, and other devices.

MapReduce—A programming model for processing large quantities of data in parallel on many nodes in a computer cluster.

MongoDB—An open source document database and the leading NoSQL database.[2]

NoSQL—A database system in which data is not stored in traditional relational form but rather in key-value, graph, or other format.

Quantified Self—A movement to better understand ourselves by tracking our personal data.[3]

Real-time—The analysis of data as it becomes available. Real-time analysis is in contrast to batch-based analysis, which occurs minutes, hours, or days after data is received.

Relational database—Software that stores data in a structured form that indicates the relationships between different data tables and elements. Oracle, Microsoft SQL Server, MySQL, and PostgreSQL are well-known relational databases.

Semi-structured data—Data that shares characteristics of both structured and unstructured data.

---

[2]http://www.mongodb.org/
[3]http://quantifiedself.com/

SQL—Structured query language; a language for storing data to and retrieving data from relational databases.

Unstructured data—Data such as raw text, images, and videos that does not contain well-defined structures defining how one piece of data relates to another.

Visualization—A way to see large quantities of data in graphical form.

Volume, variety, and velocity—also known as the three Vs, these are three commonly used measures of Big Data. Volume is how much data there is; variety refers to the kinds of data; and velocity refers to how quickly that data is moving.

Index

A

Adaptive learning system
  built-in feedback loop, 175
  district administrators, 176
  eAdvisor system, 175
  Knewton system, 175
  online learning systems, 174
  solutions, 187
Amazon Elastic MapReduce (EMR), 90
Amazon Mechanical Turk (AMT), 152
Amazon Web Services (AWS), 6, 22
  cost reduction, 90
  EMR, 90
  Glacier, 91
  Hadoop, 90
  high bandwidth connection, 88
  HPC, 90
  Kinesis, 91
  on-demand scalability, 91
  physical servers, 90
  revenue, 89
  SLA, 91
  storage and computing capacity, 89
  users, 89
  web hosting, 89
Apache Cassandra, 20
Apache Hadoop, 20
Apache HBase, 20
Apache Lucerne, 21
Application performance
        monitoring (APM) market, 118
Aspera, 24

B

Big Data
  AWS, 6
  BDA
    advantage, 11
    Hadoop software, 11
    IT administrators, 11
    mobile applications and
        recommendations, 12
    Opower, 12
    SaaS, 12
    Splunk, 11
  business users, 2
  definitions, 1
  disruption, 10
  Google search engine, 4
  health application
        (see Health application)
  information advantage, 8
  landscape (see Landscape)
  learning opportunities
        (see Learning opportunities)
  real time analysis, 13
Big Data applications (BDAs), 111
  advantages, 11, 132
  Apache Hadoop distribution, 11, 98
  application layer, 88
  AWS
    cost reduction, 90
    EMR, 90
    Glacier, 91
    Hadoop, 90
    high bandwidth connection, 88
    HPC, 90

Big Data applications (BDAs) (*cont.*)
  Kinesis, 91
  on-demand scalability, 91
  physical servers, 90
  revenue, 89
  SLA, 91
  storage and computing capacity, 89
  users, 89
  web hosting, 89
benefits, 98
billion dollar company, building
  AppDynamics, 134
  big wave riding, 133
  enterprise software products, 133
  market selction, 133
  New Relic, 134
  Oracle, 134
  out-of-the-box solution, 133
  salesforce, 134
  Software AG, 134
  Splunk, 133–134
  TIBCO, 134
  vendors, 134
  web companies, 133
business intelligence, 31
business models
  cloud-based billing company, 138
  cost predictability, 139
  dot-com crash, 140
  limited partners, 139
  service model, 138
  subscription models, 138
  usage-based software, 139
  vendors, 139
business needs, 95
core assets, 132
creation, 131
custom applications, data analysis, 28
data as a service, 32
data building, 95
data-driven approach
  Action Loop, 127–128
  confidence and conviction, 127
  data access, 127
  historical evidence, 127
  marketing investment, 127
  network servers, 127

data sources
  customer contact information, 96
  customer support requests, 96
  real estate information, 97
  semi-structured data, 96
  site performance data, 96
  unstructured data, 96
  visitor information, 96
data storage, 97
data visualization, 30
e-commerce transaction, 86
engineering team, 100–101
Facebook, 27
feedback loop
  benefits, 130
  data collection and analysis, 129
  electrical shock hurt, 129
  massive scale, 130
  outcomes, 129
  slow and time consuming, 130
Google, 27
high-speed broadband, 86
infrastructure innovation, 27
infrastructure services, 86–87
innovative data services, 137
internal data sources, 137
investment trends, 135–136
IT administrators, 11
LinkedIn, 27
low-cost sensors, 137
marketing (see Big Data marketing)
MDS, 130–131
mobile
  applications, 12, 94
  electronic health records, 93
  fitness, 93
  Google glass, 94
  IoT, 93
  logistics and transportation, 94
  platform approach, 94
  software installation, 93
Netflix, 27
noise, 128
NoSQL database, 98
online advertisement, 28
operational intelligence, 32
Opower, 12

PaaS solution, 87
Pandora, 27
pre-built modules, 99
private cloud services
    AWS, 92
    in-house hardware and software, 92
    on-demand scalability, 92
    spot instance prices, 93
project (see Big Data project)
public cloud services
    computing power, 92
    cordon off infrastructure, 92
    demand spikes, 92
    spot instance prices, 92
QlikView visualization product, 30
real estate data, 137
SaaS products, 12
sales and marketing
    CRM, 29
    data sources, 30
    data visualization, 29
    marketing automation, 29
    performance monitoring, 29
signal, 128
Splunk software, 11
SQL, 97
transaction data, 132
Twitter, 27
visualization, 99–100
web application performance data, 86
Big Data marketing
automated modeling
    content management, 149
    creative component, 147–148
    delivery component, 148–149
    marketing software, 149
    two-fold approach, 148
business value, 142
CMO
    Adobe Omniture/Google
        Analytics, 145
    brand awareness
        and purchasing, 144
    CIO, 144
    CPG, 146
    customers value, 146
    Google AdWords, 144–145
    marketing expenses, 144
    SaaS model, 145

    software solutions, 145
    spreadsheet, 145
    web-based payment system, 145
content marketing
    (see Content marketing)
conversation improvement
    analytics data, visitors, 143
    Google, 143
    media company, 144
    offline channels, 143
customer action, 146–147
execution, 142
Internet forums, 154
quants, 155
ROI, 154–155
sentiment analysis/opinion mining, 154
technical advantages, 142
vision, 142
Big Data project
churn reduction, 111
company challenges, 121
consumers benefit, 122
customer value, 123
data policy, 106
driving behavior, 121
identification, 105
improve engine efficiency, 122
marketing analytics
    APM, 118
    AppDynamics, 118
    BI systems, 118
    cloud-based approach, 118
    CRM, 118
    data gathering, 119
    email campaigns, 117
    iteration, 121
    New Relic product, 118
    results analysis, 119
mobile application, 122
OpenXC interface, 122
outcomes, 104
resources, 108
value measurements, 107
visualization, 110
workflow
    analysis, 115
    call towers and
        network capacity, 116
    hypothesis creation, 112

Big Data project (*cont.*)
    hypothesis-driven approach, 116
    program implementation, 116–117
    system set up, 113
    transformations, 114

Business intelligence (BI), 118

Business-to-business (B2B), 139

C

CabSense, 168

Call Detail Records (CDRs), 113

Centers for Disease Control (CDC), 166

Chief data officer (CDO), 45

Chief marketing officer (CMO)
    Adobe Omniture/Google Analytics, 145
    brand awareness and purchasing, 144
    CIO, 144
    CPG, 146
    customers value, 146
    Google AdWords, 144–145
    marketing expenses, 144
    SaaS model, 145
    software solutions, 145
    spreadsheet, 145
    web-based payment system, 145

Cloud-based approach, 118

Cloud computing. *See also* Big Data
    applications (BDAs)
    advantage, 23
    Amazon.com, 22
    CRM solution, 21
    customer data management, 23
    data generation, 21
    data transfer, 24
    high volume data, 23
    reliability service, 24
    SaaS model, 21

Cloudera, 25–26

Codecademy, 177

Consumer packaged goods (CPG), 146

Content marketing
    BloomReach, 150
    crowdsourcing
        AMT, 152
        blogs, 153

        Freelancer.com, 152
        high-value content, 153
        low-value content, 153
        oDesk.com, 152
        outsourcing task, 152
        SEO, 152
        TaskRabbit, 152
        webinars and webcasts, 153
    Google's search index, 149
    individual products, 150
    LinkedIn, 150
    media company, 154
    product seller, 150
    public relation, 151

Coursera, 178

Customer relationship
    management (CRM), 21, 118, 170

D

Data-driven approaches
    Action Loop, 127–128
    confidence and conviction, 127
    data access, 127
    historical evidence, 127
    marketing investment, 127
    network servers, 127

Data informs design
    apple design, 52
    architecture, 58
    business interests, 51
    car designer, 55
    classic music halls, 57
    data-driven design, 59
    Facebook applications, 51
    game design, 53
    network interests, 50
    photo uploader, 51
    qualitative data, 50
    quantitative data, 50
    reverberation time, 57
    strategic interest, 50
    web site design (see Web site design)

Data visualization
    creation
        CartoDB, 70
        desktop software application, 70
        file-based data sources, 71

HighCharts, 70
Public Data Explorer, 70
Tableau Desktop, 71–72
data investors, 67
geographic visualization, 68
image capturing and sharing, 75
Infographics, 76
Instagram, 75
key indicator managers, 67
knowledge data compression, 72–73
multiplier effect, 83–84
network diagrams, 69
pattern detection, 82
psychology and physiology, 82–83
public data sets
Common Crawl Corpus, 77
dashboards, 77
data animation, 78
data learning, 77
Nightingale's
coxcomb diagram, 78
online resources, 77
U.S. Census data, 77
real-time process
data capture, 80
data storage and analysis, 79
infographics, 79
Nielsen ratings, 79
sentiment analysis, 79
Twindex, 80
Twitter, 80
textual information, 81
time series, 69
Tufte contributions
Current Biology, 74
data communication, 74
education and life history, 73
infographics, 74
information communication, 74
multi-electrode array, 74
U.S. Census Bureau data, 66–67
U.S. Census population data, 64–66
Washington D.C.
information handling, 63
subway map, 64
Tokyo Metro and Toei, 64
word maps, 69
DreamBox, 175, 180

E

eAdvisor system, 175
EdX, 177
Elastic MapReduce (EMR), 22
Electronic Health Record (EHR), 163
Extract/transform/load (ETL) process, 20

F

Facebook
affect health and personality, 198
genuine interactions, 199
relationship status, 198
six degrees concept, 197
social graph, 196–198
worldwide users, 196

G

Global positioning system (GPS), 108
Google search engine, 4

H

Hadoop, 26
Hadoop cluster, 114
Hadoop distributed file system (HDFS), 115
Health application
BodyMedia armband, 160
data analytics company 23 and Me, 158
DNA analysis services, 159
DNA testing, 158
energy consumption, 168
financial institutions, 170
Fitbit, 160
fitness, 161
Garmin Connect services, 158
healthcare costs, 161
health data collection, 162
health information, monitoring, 162
human genes, 159
improvements, 161
individual genetic disorders, 159
Ironman athletes, 157
low-cost cloud services, 162
mobile devices, 170–171
nutrition intake, 160

Health application (cont.)
   Parkinson's disease, 162
   patient health history
      benefits, 163
      drchrono, 163
      EHR, 163
      health profiles, 164
      HITECH Act, 163
      imaging system, 164
      self-monitoring and
         health maintenance, 164
   photic sneeze reflex, 159
   PSA
      ablative surgery, 165
      access data, 166–167
      CDC, 166
      CellMiner, 166
      common cold, 167
      data and insights, 165
      hormone therapy, 165
      mortality rates, 165
      NCI, 166
      pattern recognition, 165
      prostate cancer, 164
      psychological impact, 165
      smoking and lung cancer, 166
      vaccines, 167
   public transportation, 168
   quantified self, 162
   retailers, 169
   smart city, 168
Health Information Technology for
      Economic and Clinical Health
      Act (HITECH), 163
High-performance computing (HPC), 90

I

Information Management System (IMS), 40
Internet of Things (IoT), 93

J

Jevons paradox, 10

K

Khan Academy, 176
Knewton system, 175

L

Landscape
   BDA (see Big Data applications (BDAs))
   chart, 15–16
   cloud computing
      advantage, 23
      Amazon.com, 22
      CRM solution, 21
      customer data management, 23
      data generation, 21
      data transfer, 24
      high volume data, 23
      reliability service, 24
      SaaS model, 21
   data cleansing, 33
   data privacy and security, 33
   data visualization, 16
   file sharing and collaboration, 33
   infrastructure, 16, 25
   market growth rate
      Facebook, 17
      Twitter, 18
      Walmart, 18
   open source
      Apache Hadoop, 20
      Apache Lucerne,
        text search engine, 21
      computing cost reduction, 20
      database availability, 19
      ETL process, 20
      "freemium" business models, 21
      Linux, 19
      MySQL database project, 19
      relational databases, 19
Learning opportunities
   A/B testing, 176
   adaptive learning system
      built-in feedback loop, 175
      district administrators, 176
      DreamBox, 175
      eAdvisor system, 175
      Knewton system, 175
      online learning systems, 174
      solutions, 187
   brains store information, 182
   Codecademy, 177
   cognitive overload, 182
   college dropout rate, 173

course materials, 176
data-driven approach, 174, 188
EdX, 177
effective learning, 181
grouping material, 182
hands-on environments, 179
intensive listening, 186
Khan Academy, 176
language acquisition, 184
mathskill
    English numbering system, 183
    fractions, 183
    higher earnings, 183
    mathemagics, 184
    probability, 183
    working memory loop, 183
MDM, 187
online courses
    Coursera, 178
    distance learning programs, 178
    MOOC, 178
    simulation environments, 179
    Udacity, 179
    Udemy, 178
    virtual environments, 179
Pandora, 174
pattern matching, 186
semantic filtering, 187
semantic search technology, 187
track performance
    DreamBox, 180
    faculty-student ratios, 180
    MOOC, 180
    public school performance, 180

M

MapReduce, 20
Massive open online
    course (MOOC), 178, 180
Master Data Management (MDM), 187
Match.com, 192
Mental map-making, 195
Minimum Data Scale (MDS), 130–131
Multiple high-profile consumer companies, 2

N

National Cancer Institute (NCI), 166
Netflix, 24, 27
Net Promoter Score (NPS), 104

O

OkCupid, 190
Online dating and
    social networking
calendar-based applications, 189
Facebook
    affect health and personality, 198
    genuine interactions, 199
    relationship status, 198
    six degrees, 197
    social graph, 196–198
    worldwide users, 196
flirty, smiling, and not smiling, 190
lies, damn lies and statistics, 192
marriage success prediction, 194
missing data finding, 193
mobile applications, 189
OkCupid, 190
romance
    Match.com, 202
    relationship, 203
    Skout, 202
    smartphones, 202
    virtual gifts, 201
social capital
    bonding, 199
    bridging, 199
    broadcasting, 200
    definition, 199
    direct communication, 200
    It's Just Lunch, 201
    one-to-one
        communication, 200–201
    online identity, 201
    passive consumption, 200
    personal relationship, 200
OpenXC interface, 122
Organisation for Economic Co-operation
    and Development (OECD), 173

P, Q

Platform as a Service (PaaS) solution, 87

Prostate-specific antigen (PSA)
    ablative surgery, 165
    access data, 166–167
    CDC, 166
    CellMiner, 166
    common cold, 167
    data and insights, 165
    hormone therapy, 165
    mortality rates, 165
    NCI, 166
    pattern recognition, 165
    prostate cancer, 164
    psychological impact, 165
    smoking and lung cancer, 166
    vaccines, 167

R

Relational database management
    system (RDBMS), 19, 40

Return on investment (ROI), 154

Roadmap
    analysis, 38
    automated interpretation, 40
    CDO, 45
    compile and query, 37
    data scientist and analyst, 44
    geographic visualizations, 40
    hardware, 42
    NoSQL, 36
    RDBMS, 40
    system administrators, 44
    table-based data store and SQL, 36
    vision, 47

S

Search engine optimization (SEO), 152

Semantic search technology, 187

Service level availability (SLA), 91

Skout, 202

Social capital
    bonding, 199
    bridging, 199
    broadcasting, 200
    definition, 199
    direct communication, 200
    It's Just Lunch, 201
    one-to-one communication, 200–201
    online identity, 201
    passive consumption, 200
    personal relationship, 200

Social graph, 196–197

Software as a Service
    (SaaS) model, 12, 21, 145, 170

Solaris, 19

Structured query language (SQL), 36, 97

T

Tableau Desktop, 71, 115

U, V

Udacity, 179

Udemy, 178

Universal Coordinated Time (UTC), 37

W, X, Y, Z

Web site design
    A/B tests, 60, 62
    analytics tools, 60
    content management systems, 61
    data-driven design, 62
    design elements, 61
    Google's recent
        algorithm changes, 62
    limitations, 61
    mobile applications, 62
    time-consuming and
        difficult process, 61
    Wix, 60

Get the eBook for only $10!

Now you can take the weightless companion with you anywhere, anytime. Your purchase of this book entitles you to 3 electronic versions for only $10.

This Apress title will prove so indispensible that you'll want to carry it with you everywhere, which is why we are offering the eBook in 3 formats for only $10 if you have already purchased the print book.

Convenient and fully searchable, the PDF version enables you to easily find and copy code—or perform examples by quickly toggling between instructions and applications. The MOBI format is ideal for your Kindle, while the ePUB can be utilized on a variety of mobile devices.

Go to www.apress.com/promo/tendollars to purchase your companion eBook.

Other Apress Business Titles You Will Find Useful

Supplier Relationship Management
Schuh/Strohmer/Easton/
Hales/Triplat
978-1-4302-6259-6

Why Startups Fail
Feinleib
978-1-4302-4140-9

Disaster Recovery, Crisis Response, and Business Continuity
Watters
978-1-4302-6406-4

Healthcare Information Privacy and Security
Robichau
978-1-4302-6676-1

Eliminating Waste in Business
Orr/Orr
978-1-4302-6088-2

Data Scientists at Work
Gutierrez
978-1-4302-6598-6

Exporting
Delaney
978-1-4302-5791-2

Unite the Tribes, 2nd Edition
Duncan
978-1-4302-5872-8

Corporate Tax Reform
Sullivan
978-1-4302-3927-7

Available at www.apress.com